Constrained Deformation of Materials

Y.-L. Shen

Constrained Deformation of Materials

Devices, Heterogeneous Structures and Thermo-Mechanical Modeling

 Springer

Y.-L. Shen
University of New Mexico
Dept. Mechanical Engineering
Albuquerque, New Mexico 87131
USA

ISBN 978-1-4419-6311-6   e-ISBN 978-1-4419-6312-3
DOI 10.1007/978-1-4419-6312-3
Springer New York Dordrecht Heidelberg London

Library of Congress Control Number: 2010931376

Printed on acid-free paper

Springer is part of Springer Science+Business Media (www.springer.com)

Preface

The fast advancement of materials and device technologies has necessitated research-ers and engineers to devote greater attention to material behavior at small scales. At the center of attention is mechanical deformation, which not only directly controls the thermo-mechanical performance but also exerts influences on other increasingly sophisticated functionalities. In miniaturized structures, deformation is heavily affected by the physical confinement imposed on the material. In materials possessing distinct micro- and nano-scale heterogeneities, the mismatching phases engender internal constraint within the material itself. As a consequence, mechanical deforma-tion in its simplistic form can no longer occur. Rather, deformation takes place in a constrained manner when subjected to mechanical loading and temperature excur-sion, leading to complex deformation patterns which can have tremendous impacts on the performance and reliability of these device and materials systems.

This book is written to address the constrained deformation phenomena frequently encountered in modern engineering. The coverage encompasses thin films, microelec-tronic devices and packages, as well as composite and multi-phase materials. A unique attribute of the book is that it brings out the commonality of these diverse topics and threads them together under a unified theme. The focus is largely on applying numeri-cal modeling at the micro-mechanical level and its connection to physical material behavior, augmented by a multitude of case analyses. Attention is also devoted to the engineering relevance so implications to real-life reliability issues can be examined in detail. To provide an integrated treatment, the continuum mechanics-based phenome-nological approach is adopted. However, contact with the materials science aspects of constrained deformation is still made throughout the chapters, wherever appropriate.

This book advocates the proper application of engineering tools, such as com-mercial or in-house finite element analysis programs, to aid in mechanistic under-standing as well as interpretation of physical features and experimental observations. In these regards, I have attempted to illustrate that simple models could indeed go a long way. Efforts were made to keep the descriptions of theories and mathemati-cal expressions to a minimum, certainly no more than what is needed to understand the context or to develop the skill for undertaking similar kind of deformation analyses. Both seasoned and would-be numerical modelers should find the book valuable. In addition, experimentalists can benefit from the mechanics insight provided throughout the discussion.

The first chapter presents the scope of the book and sets the scene for subsequent developments. The necessary background on solid mechanics and materials modeling are given in Chap. 2. Chapters 3 through 6 are devoted to the individual topics of constrained deformation in thin films, patterned film structure in microelectronic devices, electronic packaging structures, and heterogeneous materials. The treatment in each of these chapters is self-sufficient. It is recognized that each of these topics could itself be expanded into a sizable volume. Thus I had to rely on my own judgment in selecting the subtopics for a comprehensive and coherent coverage. Finally, the book is concluded with Chap. 7 which remarks on salient challenges and directions of future endeavor. The total number of references cited at the end of each chapter amounts to over 500. They serve as useful sources for further information or more in-depth studies.

The book is designed for researchers and engineers in academia and industry, as well as students. While a research monograph in nature, it can also be used as a textbook or reference book at the graduate and senior undergraduate levels. Specialized topical courses on, for instance, deformation analysis of materials, mechanics of thin films and devices, micro-mechanical modeling, and applications of the finite element method will be suitable. A distinctive feature of the book is that, in each of Chapters 3 through 6, there is a last section titled "Projects" which suggests numerous broad modeling ideas. They are built upon the contents in the previous sections, and may serve as homework or term projects or simply as impetuses for further exploration. Many of the projects can also be expanded to become useful thesis research topics.

I have assumed that readers have had previous exposure in introductory materials science, and possess some basic knowledge of solid mechanics and finite element modeling. Background on computational mechanics theories, however, is not required. In all the modeling examples, details of the finite element implementation (discretization, numerical algorithm, etc.) are left out from the presentation. Rather, the application and physical aspects of the problems are emphasized. Throughout the text I use the material names and their chemical symbols interchangeably. In addition, there is no differentiation between the terms "modeling" and "simulation." Careful readers may also find that some mechanical properties for a given material, used as the model input, are not exactly the same when they appear in different sections. This is due to the slight inconsistency of material parameters taken from different sources, and to the fact that certain thermo-mechanical properties of nominally the same material are a function of its precise composition, microstructure and processing history.

The concept for the book grew out of my nearly 20 years of research experience in this area. I am grateful to many of my mentors, colleagues and students who have kept me motivated over the years. Several individuals particularly deserve my mention: S. Suresh, A. Needleman, N. Chawla, A. Mortensen, and I. A. Blech. I would also like to acknowledge National Science Foundation, Sandia National Laboratories and Air Force Office of Scientific Research for sponsoring some of my research work toward the development of this book.

Finally, I wish to express my profound gratitude to my family for their ever-present support, encouragement and understanding.

Contents

Chapter 1
Introduction

When an external load is applied to a material, deformation occurs. The concept of "rigid body," while very useful in a wide range of structural and dynamic engineering applications, is never a true representation of actual materials. This non-rigidity of material gives rise to a dimensional change (strain) in response to the outside mechanical stimulation. If instead it is the dimensional change that is prescribed, then a mechanical stress will be generated in the material. The relationship between stress and strain determines the constitutive response of a material. The simplest form of stress–strain relationship is linear elastic, where the state of strain can be directly obtained from a known state of stress (and vice versa) through a set of elastic constants. More complex behavior such as plastic and time-dependent deformation frequently occurs, depending on the combination of materials, geometry, and loading conditions.

Traditionally, engineering components are often designed to ensure loading within a stress or strain limit. With complex geometric and/or loading configurations, deformation analyses have to be performed before deciding if the limit will be reached at any specified location. For materials containing a heterogeneous internal structure, even a simple loading mode can create a highly disturbed deformation field. In recent years the development of increasingly miniaturized functional devices and structural components has led to an expanded interest in mechanical deformation of materials at small scales. Applications of immediate technological importance include surface coatings, microelectronic devices and packages, microelectromechanical systems (MEMS), specialized composites and nanostructures, among others. These miniaturized structures are almost always bonded to one or more external media. Under mechanical loading, deformation in any single component of the structure does not take place in an unrestricted form, because it will be influenced by the adjoining component(s). Therefore deformation occurs in a *constrained* manner. If a temperature change is involved, the mismatch of coefficients of thermal expansion among different materials will also cause deformation which is inherently constrained by the adjoining material(s). In many cases excessive deformation is a precursor to failure, so it directly relates to the reliability and performance of the device or component. In other applications, however, deformation is intentionally induced and controlled so as to provide special functionalities.

Y.-L. Shen, *Constrained Deformation of Materials: Devices, Heterogeneous Structures and Thermo-Mechanical Modeling*, DOI 10.1007/978-1-4419-6312-3_1,
© Springer Science+Business Media, LLC 2010

It is therefore of paramount importance to be able to quantify constrained deformation for the purposes of design and reliability assessment, as well as for gaining fundamental insight into the physical processes.

This book offers readers a basic understanding of the problems and introduces the important concepts and approaches to deal with them. A great number of case studies encompassing prototypical and real-life small structures and devices are included. We emphasize the versatility of employing *simple engineering tools* for undertaking the deformation analysis. Although the discussion appears to center around modeling and simulation, one essential goal is also to provide guidance for designing physical tests and to aid in the proper interpretation of experimental results.

1.1 Simple Illustration of Constrained Deformation

As an introductory example to illustrate the elementary features, we consider a simple square shaped homogeneous elastic solid. The material is undergoing uniform stretching along the y-direction as shown in Fig. 1.1a. For simplicity the material is assumed to be a two-dimensional (2D) domain under the plane strain condition (with no displacement allowed in the direction perpendicular to paper). Figure 1.1a is in fact a contour plot obtained from a finite element analysis, showing the distribution of tensile strain ε_{yy} inside the material when the applied tensile strain is 0.001. Of course, the result in Fig. 1.1a is a trivial one since the deformation is not constrained in any in-plane direction so a uniform shading is seen. Figure 1.1b–d, however, show more complex patterns under the same applied overall strain as explained below.

In Fig. 1.1b, two fibers are embedded within a matrix having the same properties as the homogeneous material in part (a). Here the fiber material is much stiffer than the matrix, with all other conditions remaining unchanged. It is seen in Fig. 1.1b that the deformation field becomes highly non-uniform. Under the same global (macroscopic) strain, the local strain can be high in certain regions and low elsewhere. Deformation thus occurs in a *constrained* manner. If the material is physically confined from the outside, such as in Fig. 1.1c where the bottom boundary is rigidly clamped, local deformation is disturbed even in a homogeneous material. In Fig. 1.1c the material is under the pulling action which results in the same dimensional change along the y-direction as in parts (a) and (b). In (c) the lateral contraction in response to the pulling is prohibited along the clamped boundary, leading to a complicated strain field inside the homogeneous material. Figure 1.1d shows the case which is a combination of those in parts (b) and (c): two fibers are embedded within the matrix which is also rigidly clamped at the bottom. The complexity of deformation pattern due to this combined internal and external constraint is evident.

While the illustration in Fig. 1.1 considers a simple form of mechanical loading, it is recognized that thermal loading (involving temperature changes) will also result in constrained deformation in the cases of (b), (c) and (d), albeit in a different manner. Further, a greater degree of intricacy will be involved if the inelastic response of the material, such as plastic yielding, starts to play a role.

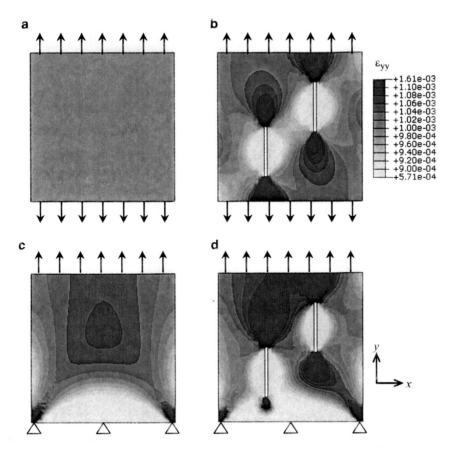

Fig. 1.1 Simple finite element analysis illustrating the disturbance of deformation caused by physical constraint of the material. Contour plots of the strain component ε_{yy} are shown when the applied stretching results in a nominal macroscopic strain of 0.001 along the y-direction for the cases of (**a**) a homogenous material (pure matrix) under uniaxial stretching, (**b**) the same matrix with two stiff fibers embedded within, (**c**) the pure matrix with its bottom boundary fixed in space, and (**d**) the same matrix with two stiff fibers embedded within and with the bottom boundary fixed in space. Both the matrix and fiber materials are linear elastic, with Young's modulus and Poisson's ratio of the matrix being 70 GPa and 0.33, respectively, and Young's modulus and Poisson's ratio of the fiber being 450 GPa and 0.17, respectively. The two-dimensional plane strain condition is assumed. In (**b**) and (**d**), only the strains inside the matrix are shown

In this book we will devote our attention to many of these and relevant issues frequently encountered in miniaturized structures and materials.

1.2 Applications

In this section we present a few examples of actual materials or devices, the performance and reliability of which are directly influenced by the constrained deformation. Emphasis is placed on direct technological relevance of the subject matter.

It is noted that the examples given here only represent a small subset of the rich application areas of micro- and nano-engineered systems.

1.2.1 Deformation in Micromachined Structures

MEMS, also known as micromachines, have become an integral part of modern engineering. Building on the advancement of microelectronics manufacturing technologies over the past decades, MEMS can now be produced to incorporate mechanical, electrical, optical, chemical and/or biological functions into devices at very small scales. Materials related issues have always played a central role in their development [1–3]. MEMS-based techniques can also be utilized to characterize material properties. This is particularly valuable because the mechanical properties of materials with small dimensions often depend on their physical size and processing route. An integrated approach of building the material into a miniaturized test structure thus offers great advantages.

Figure 1.2 shows a scanning electron microscopy (SEM) photograph of several micromachined cantilever beams [4]. Each beam consists of a silicon nitride thin film on top of a silicon oxide thin film, each less than 0.5-μm thick. Therefore, deformation is not only constrained by the clamped end but is also influenced by the mutual constraint of the two materials. In the figure there is no external loading applied to the beams; the downward bending is caused by the mismatch in residual stresses between the two films. If a temperature difference is applied to the same

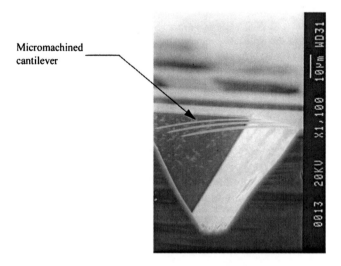

Micromachined
cantilever

Fig. 1.2 SEM photograph showing the silicon nitride/silicon oxide bi-layer cantilever beams produced by bulk micromachining [4]. The mismatch in residual stresses in the two thin films leads to bending (Reprinted with permission from IOP Publishing Limited)

Fig. 1.3 Cantilever beams of tantalum produced by micromachining [5]. Bending of the beam is a consequence of the gradient of residual stress along the thickness of the thin film (Reprinted with permission from Materials Research Society)

structure, further bending will occur due to the mismatch in thermal expansion coefficient of the two materials.

Figure 1.3 shows another example of micromachined cantilever beams [5]. In this case the beam consists of only one material, 1-μm thick tantalum (Ta). Although any mismatch effect is non-existent for the single material, the gradient of the residual stress across the thickness direction still results in a bending profile. In the figure the upper part of the tantalum film carried a stress which is more tensile than the lower part.

Similar types of the beam structure can also be produced and subject to mechanical bending. Many of the deformation features will be discussed later in the book.

1.2.2 Microelectronic Devices and Packages

Mechanical deformation plays a critical role in determining the robustness and reliability of microelectronic devices and packages. Figure 1.4 shows a cross section of the interconnect structure in a microprocessor [6]. The lighter areas are mainly traces of thin copper lines (line direction is perpendicular to paper), some connected by vertical vias, above the transistors on a silicon chip. The lines are used for transmitting electric signals among the transistors themselves and between transistors and the external circuitry at the packaging level. The thicknesses of the metal lines shown in the figure are approximately from 100 (lower level) to 500 nm (upper level). The darker regions are mainly carbon-doped silicon oxide dielectrics. There are also very thin barrier layers between copper and the dielectric, thus forming

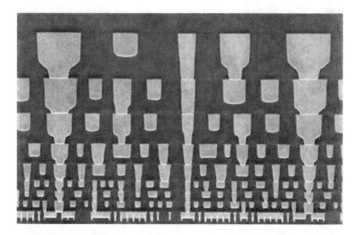

Fig. 1.4 Cross section of multilevel interconnects in a microprocessor [6]. The copper lines, with approximate thicknesses from 100 to 500 nm, are perpendicular to the paper. Lines at different levels are connected by vertical vias, and the metal structure is embedded within dielectric materials (Photograph © 2008 IEEE; reprinted with permission)

an extremely complex composite structure in small dimensions. Stresses induced by differential thermal expansion can lead to interconnect voiding damage, interfacial delamination, and dielectric cracking etc. during manufacture and service. Stresses in the conductor lines will also influence electromigration, which occurs as a result of interaction between diffusing atoms and mobile electrons. Deformation of the external package structure is another important source of stresses that can be experienced by the silicon chip.

A sample electronic packaging structure is shown in Fig. 1.5. It is a package of power electronics where the primary function is to control or convert electrical energy. In the figure a cross section of a portion of the package, used in typical adjustable-speed motor drives or moderate-power radio-frequency amplifiers, is shown [7]. Here the silicon chip is soldered on a substrate consisting of alumina clad on both sides with a patterned copper layer. The top side of silicon is connected to gold or aluminum wire bonds and dielectric; the underside of the substrate is attached to a copper base plate which is in turn attached to a heat sink assembly. In this case, the primary thermo-mechanical reliability concern still originates from the thermal expansion mismatches among the different materials. The development of stress field in the complex structure depends strongly on the materials and their detailed geometry. Instantaneous mechanical failure may occur as a result of the stress. Repeated temperature cycles can also lead to nucleation and propagation of fatigue cracks. In addition, time-dependent deformation can lead to creep failure, especially in the low-melting-point solder material.

The power electronics package described above possesses a physical dimension on the order of tens of millimeters. In many other types of device packages, the reliability demand is even more stringent. One such example is the ball grid arrays, where tens to hundreds of individual solder joints exist and the feature dimension

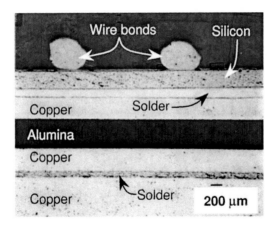

Fig. 1.5 Cross section of a package structure in power electronics [7]. The silicon chip is soldered on a copper-alumina-copper substrate, which is in turn soldered on a copper base plate attached to the heat sink. Wire bonds embedded in the dielectric material above silicon can also be seen (Reproduced by permission of the MRS Bulletin)

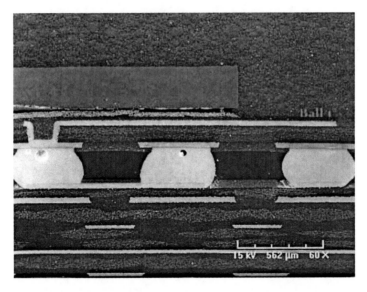

Fig. 1.6 Cross section showing a chip scale package and circuit board located, respectively, above and below the solder joints [8]. Other segments of metal traces and pads can also be seen (Photograph © 2008 IEEE; reprinted with permission)

of individual joints is much smaller (e.g., tens to hundreds of microns). Failure of a single joint may signify failure of the package. A representative structure used in a test vehicle for drop impact is shown in Fig. 1.6 [8]. The cross sections of three solder joints are in the figure. Some copper pads and traces are also visible as lighter regions. The solder joints are used to connect the chip scale package (upper side) and the circuit board (lower side). Note that the solder joint is made out of soft low-melting-point alloys so they are especially prone to deformation. For instance, when drop impact occurs to mobile or handheld consumer electronics, fluctuating high-rate deformation will be experienced by the solder joints inside. Cracking in the solder alloy or along the interface may occur. In fact, even in the absence of mechanical load, the joints still undergo slow cyclic deformation because the components they connect possess thermal expansion mismatch during regular on–off temperature periods. This form of deformation eventually results in initiation and propagation of fatigue cracks, which has long been regarded as a major reliability threat in microelectronic packages. Note that deformation in solder, no matter of mechanical or thermal origins, is primarily transmitted through the adjacent materials. A highly constrained manner is to be expected, as will be discussed in a later chapter.

1.2.3 Materials with Internal Structure

All materials possess internal heterogeneity to a certain degree. The apparent examples are multi-phase materials and composite materials. Even if they are not bonded to an external structure, deformation in the material will be far from uniform. Figure 1.7 shows a scanning electron microscopy picture of a tin (Sn)–lead (Pb) alloy with a near eutectic composition (60Sn–40Pb). The Sn-rich solid solution and Pb-rich solid solution constitute the dual phase structure. Depending on the processing details (e.g., different cooling rates from the molten state), the phase distribution can be more lamellar or more globular. When the material is subject to loading, deformation is internally constrained due to the mismatch of mechanical

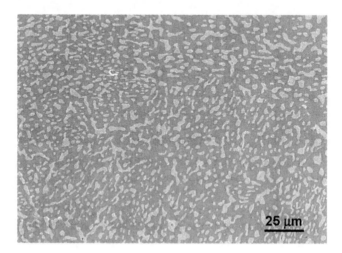

Fig. 1.7 Scanning electron micrograph showing the microstructure of a near eutectic tin (Sn)-lead (Pb) alloy. The Sn-rich and Pb-rich phases are the *darker* and *lighter regions*, respectively, in the picture

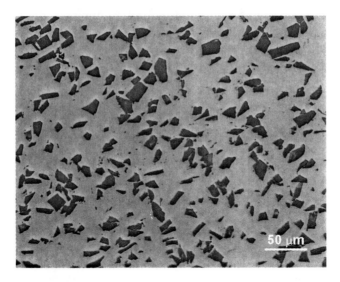

Fig. 1.8 Optical micrograph showing an aluminum alloy (*light matrix*) reinforced with silicon carbide particles (*dark regions*)

properties between the phases. (In this particular alloy the interface sliding characteristic also affects the overall mechanical response [9].)

Figure 1.8 shows an optical micrograph of a type of metal matrix composites: aluminum (Al) alloy reinforced with silicon carbide (SiC) particles. From a mechanical standpoint, the composite is equivalent to a multi-phase material. In the present case the matrix phase and the reinforcement have distinctly different properties. Deformation in the ductile aluminum matrix is severely constrained by the hard ceramic particles. A similar situation also exists if the reinforcement is of different geometric forms.

1.3 Outline and Scope

As mentioned above, the use of simple modeling tools and proper application of them are emphasized. Common misconceptions for the interpretation of experimental results will also be discussed throughout the book.

In Chap. 2, an overview of the basic concepts of engineering solid mechanics is given. The presentation focuses on continuum-level descriptions of the elastic, elastic-plastic and time-dependent material behaviors, which are of direct relevance to the deformation analysis utilized in this book. A brief introduction to modeling techniques, as well as useful skills and "tips" for conducting the deformation analysis, are also included. It is understood that some sub-continuum simulation methods, including atomistic and mesoscopic approaches, may offer more localized physical pictures and are particularly useful in certain circumstances for materials with small dimensions. However, these methods inevitably involve many idealizations in the initial, boundary and temporal conditions. Therefore, for the purpose of gaining an overall understanding and global view of deformation pattern along with measurable material responses, one still relies on the continuum-based approach [10].

Subsequent chapters are organized according to the main topics of discussion, namely the different geometric or functional features of the material. Chapter 3 is devoted to continuous thin films where the main source of deformation constraint comes from the underlying substrate material. Patterned thin films in micro-devices are the focus of Chap. 4, with special emphasis given to the interconnect structure in microelectronics. Here constrained deformation intimately influences the thermomechanical reliability of devices. This fact also applies to the electronic packages, which is the topic of Chap. 5. Chapters 3, 4, and 5 thus deal basically with *externally* constrained deformation, where the material of concern belongs to, and is thus confined by, a larger-scale outside structure. Chapter 6 focuses on heterogeneous materials, namely composite and multi-phase materials. In this case the material itself contains a structure within, so *internally* constrained deformation takes place under mechanical and thermal loading. Each of these chapters contains case studies arranged in a progressive manner, emphasizing breaking down of complicated systems into sub-structures for efficient and accurate analysis. A section named "Projects" is included at the end of each of these chapters, giving many numerical simulation ideas which are built upon the regular discussion in the text. These exercises may serve as homework or term projects, or simply as some interesting points worthy of further thinking. The projects may also be expanded to become useful thesis topics for more in-depth research. The book ends with Chap. 7 which addresses the outlook, remaining challenges as well as future directions, with regard to issues related to constrained deformation analyses.

References

1. S. M. Spearing (2000) "Materials issues in microelectromechanical systems (MEMS)," Acta Mater, vol. 48, pp. 179–196.
2. A. D. Romig, Jr., M. T. Dugger and P. J. McWhorter (2003) "Materials issues in microelectromechanical devices: science, engineering, manufacturability and reliability," Acta Mater, vol. 51, pp. 5837–5866.
3. S. Sedky (2006) Post-processing techniques for integrated MEMS, Artech House, Boston.
4. W. Fang (1999) "Determination of the elastic modulus of thin film materials using self-deformed micromachined cantilevers," J Micromech Microeng, vol. 9, pp. 230–235.
5. S. Sedky, P. Fiorini, A. Witvrouw and K. Baert (2002) "Sputtered tantalum as a structural material for surface micromachined RF switches," in R. P. Manginell, J. T. Borenstein, L. P. Lee and P. J. Hesketh: BioMEMS and Bionanotechnology, Materials Research Society Symposium Proceedings, vol. 729, pp. 89–94, Materials Research Society, Warrendale, PA.
6. S. Natarajan, M. Armstrong, M. Bost et al. (2008) "A 32 nm logic technology featuring 2nd-generation high-k + metal gate transistors, enhanced channel strain and 0.171 mm SRAM cell size in a 291 Mb array," in Proceedings of 2008 International Electron Devices Meeting, IEEE.
7. M. C. Shaw (2003) "High-performance packaging of power electronics," MRS Bull, vol. 28(1), pp. 41–50.

8. A. Farris, J. Pan, A. Liddicoat, B. J. Toleno, D. Maslyk, D. Shangguan, J. Bath, D. Willie and D. A. Geiger (2008) "Drop test reliability of lead-free chip scale packages," in Proceedings of 2008 Electronic Components and Technology Conference, IEEE, pp. 1173–1180.

9. K. C. R. Abell and Y.-L. Shen (2002) "Deformation induced phase rearrangement in near eutectic tin-lead alloy," Acta Mater, vol. 50, pp. 3191–3202.

10. Y.-L. Shen (2008) "Externally constrained plastic flow in miniaturized metallic structures: A continuum-based approach to thin films, lines, and joints," Prog Mater Sci, vol. 53, pp. 838–891.

Chapter 2
Mechanics Preliminaries

Mechanics is the "language" used for describing the mechanical behavior of materials. An understanding of basic constitutive models, such as elasticity and plasticity, is essential for undertaking deformation analyses. In this chapter the fundamental concepts and field equations are reviewed. The contents are limited to topics that are within the scope of this book and are intentionally terse. Readers who seek detailed information on solid mechanics may consult specialized textbooks (e.g., [1–7]). Other relatively brief overviews may also be found in the chapters of Refs. [8–10]. Issues concerning implementation of numerical modeling will also be addressed toward the end of the chapter.

In the presentation below tensors with the Cartesian basis, the standard indicial notation, and the small strain approximation are adopted. Attempts are made to focus only on the most commonly employed constitutive material models.

2.1 Stress and Strain

Figure 2.1 shows an infinitesimal volume element in a solid. The traction vectors (force components divided by the area of surface acted upon by the force) are shown, which also defines the stress state at this material element. Note the scalar components of the stress in Fig. 2.1 are represented by σ_{ij}, with $i, j = 1, 2$ and 3. The indices i and j denote, respectively, the vector component of the traction and the outward normal of the plane. The full stress tensor can be represented by the 3×3 matrix,

$$[\sigma_{ij}] = \begin{bmatrix} \sigma_{11} & \sigma_{12} & \sigma_{13} \\ \sigma_{21} & \sigma_{22} & \sigma_{23} \\ \sigma_{31} & \sigma_{32} & \sigma_{33} \end{bmatrix} \qquad (2.1)$$

In (2.1) σ_{11}, σ_{22} and σ_{33} are the *normal* stress components and the others are the *shear* components. Mechanical equilibrium of the material element requires that $\sigma_{ij} = \sigma_{ji}$, due to the balance of angular momentum, and

$$\frac{\partial \sigma_{1i}}{\partial x_1} + \frac{\partial \sigma_{2i}}{\partial x_2} + \frac{\partial \sigma_{3i}}{\partial x_3} + b_i = 0, \quad i = 1,2,3, \qquad (2.2)$$

Y.-L. Shen, *Constrained Deformation of Materials: Devices, Heterogeneous Structures and Thermo-Mechanical Modeling*, DOI 10.1007/978-1-4419-6312-3_2, © Springer Science+Business Media, LLC 2010

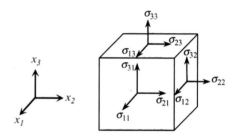

due to the balance of linear momentum, where b_i is the ith component of the body force per volume. Note (2.2) is commonly referred to as the equilibrium equations. In the absence of body force it can be written as

$$\frac{\partial \sigma_{ij}}{\partial x_i} = 0, \tag{2.3}$$

where the summation convention in standard continuum mechanics (in this case summation over i) is followed.

At every point in a stressed material there exist three mutually perpendicular directions, such that the shear components of the stress tensor always vanish when these three directions are taken as the coordinate basis. The three directions are termed principal directions and the associated normal stresses are termed *principal stresses* (denoted as σ_I, σ_{II} and σ_{III} here). The *maximum principal stress* is frequently used as a measure of the greatest normal stress value carried by a material element in a deformed structure. The following combinations of the principal stresses,

$$I_1 = \sigma_I + \sigma_{II} + \sigma_{III}$$
$$I_2 = -\left(\sigma_I \sigma_{II} + \sigma_{II} \sigma_{III} + \sigma_{III} \sigma_{II}\right), \tag{2.4}$$
$$I_3 = \sigma_I \sigma_{II} \sigma_{III}$$

are commonly referred to as, respectively, the first, second and third stress invariants. They are invariant under any coordinate transformation.

We now turn to strain. Strain is induced in a material when a physical action (e.g., stress or temperature change) causes a change in its configuration. The components of infinitesimal strain tensor are defined to be

$$\varepsilon_{ij} = \frac{1}{2}\left(\frac{\partial u_i}{\partial x_j} + \frac{\partial u_j}{\partial x_i}\right) \tag{2.5}$$

where u is the displacement vector. Note $\varepsilon_{ij} = \varepsilon_{ji}$, so the strain tensor can also be represented by a symmetric 3×3 matrix. As in the case of stress, the normal and shear components of strain are characterized by $i = j$ and $i \neq j$, respectively. It is important to pay attention to the fact that the shear component defined in (2.5) is different from the shear strain γ_{ij} typically used in engineering; they are related by $\gamma_{ij} = 2\varepsilon_{ij}$. In a deformed material, the fractional volume change ($\Delta V / V$), or dilatational strain, is given by

$$\frac{\Delta V}{V} = \varepsilon_{ii} = \varepsilon_{11} + \varepsilon_{22} + \varepsilon_{33}. \tag{2.6}$$

For an incompressible (volume-conservative) deformation, $\varepsilon_{ii} = 0$.

In many circumstances a three-dimensional (3D) deformation state may be approximated by a two-dimensional (2D) one. Here we take the $x_1 x_2$ plane to represent the 2D state so the partial differentiation of any field quantity with respect to x_3 vanishes. For the state of *plane stress*, $\sigma_{13} = \sigma_{23} = \sigma_{33} = 0$. A typical example is in a stressed thin sheet where the thickness along the x_3 direction is small. The state of *plane strain* is characterized by the condition $u_3 = 0$ and thus $\varepsilon_{13} = \varepsilon_{23} = \varepsilon_{33} = 0$. A situation where the plane strain condition prevails is inside a thick structure, where deformation in a local region is constrained along the x_3 direction by its surroundings due to certain combinations of material properties and/or load distribution.

2.2 Elastic Deformation

Most materials undergo elastic deformation when subject to a sufficiently small strain. The most simple (and commonly encountered) form is linear elasticity, where the stress and strain tensors are related by the Hooke's law,

$$\sigma_{ij} = C_{ijkl}\varepsilon_{kl} \tag{2.7}$$

where C_{ijkl} is the elastic constant tensor. For isotropic materials, the stress–strain relationship is given by the following form,

$$\varepsilon_{11} = \frac{1}{E}\left[\sigma_{11} - v(\sigma_{22} + \sigma_{33})\right]$$

$$\varepsilon_{22} = \frac{1}{E}\left[\sigma_{22} - v(\sigma_{11} + \sigma_{33})\right]$$

$$\varepsilon_{33} = \frac{1}{E}\left[\sigma_{33} - v(\sigma_{11} + \sigma_{22})\right]$$

$$\varepsilon_{12} = \frac{1}{2G}\sigma_{12} \tag{2.8}$$

$$\varepsilon_{23} = \frac{1}{2G}\sigma_{23}$$

$$\varepsilon_{13} = \frac{1}{2G}\sigma_{13}$$

where E, v and G are Young's modulus, Poisson's ratio and shear modulus, respectively. Another commonly used elastic constant is bulk modulus B, which is defined by

$$B = \frac{\sigma_H}{\Delta V / V} = \frac{\sigma_H}{\varepsilon_{11} + \varepsilon_{22} + \varepsilon_{33}} \qquad (2.9)$$

where σ_H is the hydrostatic (or mean) stress defined by

$$\sigma_H = \frac{1}{3}\sigma_{ii} = \frac{1}{3}\left(\sigma_{11} + \sigma_{22} + \sigma_{33}\right). \qquad (2.10)$$

The elastic constants introduced above are intrinsic material properties. For isotropic materials only two of the four elastic constants are independent. The constants G and B can be expressed in terms of E and v as

$$G = \frac{E}{2(1+v)} \quad \text{and} \quad B = \frac{E}{3(1-2v)}. \qquad (2.11)$$

When the material element deforms elastically by a strain increment $d\varepsilon_{ij}$, the external work is stored in the material as the strain energy. The strain energy density increment is given by $dw = \sigma_{ij}d\varepsilon_{ij}$, the integration of which over a finite strain yields the total elastic strain energy per unit volume expressed as

$$w = \frac{1}{2}\left[\sigma_{11}\varepsilon_{11} + \sigma_{22}\varepsilon_{22} + \sigma_{33}\varepsilon_{33} + 2\sigma_{12}\varepsilon_{12} + 2\sigma_{23}\varepsilon_{23} + 2\sigma_{13}\varepsilon_{13}\right]. \qquad (2.12)$$

The elastic strain energy will be completely released upon removal of the applied load. The process of removing the applied load is termed unloading.

2.3 Plastic Deformation

2.3.1 Uniaxial Response

When the stress carried by the material reaches its elastic limit, plastic deformation (or yielding) starts to occur. This is a form of deformation that cannot be recovered after unloading. The constitutive response of ductile materials such as a metal can be described as elastic-plastic. Under uniaxial loading (such as during a tensile test) the stress–strain curve follows the shape depicted in Fig. 2.2. The slope of the initial linear portion is Young's modulus (E) of the material. The yield strength, σ_y (also a material property), is practically the elastic limit of the material. For an *elastic-perfectly plastic* solid, the stress–strain curve becomes horizontal upon plastic yielding. In actual ductile materials, however, a *strain hardening* behavior exists as shown in Fig. 2.2 where the flow stress continues to increase with strain after yielding. A simple elastic-power law plastic form describing the shape of the stress (σ)–strain (ε) curve is,

$$\varepsilon = \frac{\sigma}{E} + \left(\frac{\sigma}{K}\right)^{1/n} \qquad (2.13)$$

Fig. 2.2 Schematic showing the uniaxial stress (σ)–strain (ε) curve of an elastic-plastic ductile material. The symbols E, σ_y, ε^e and ε^p represent Young's modulus, yield strength, elastic strain and plastic strain, respectively

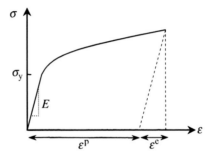

which is based on the Ramberg–Osgood relationship [11]. Here E is Young's modulus, K is a material constant, and n is the strain hardening exponent. The typical range of n for elastic-plastic metallic materials is from 0 to 0.5. Note this is only a convenient empirical equation. Other functional forms may better fit the measured response for a specific material.

With plastic deformation the total strain consists of an elastic (recoverable) part ε^e and a plastic (permanent) part ε^p. They are defined in Fig. 2.2 according to their recoverability after the stress is reduced to zero. Note that the stress–strain response during stress reduction is an elastic unloading process with a slope equal to Young's modulus of the material. If the stress continues its path into the compressive regime, reversed yielding will occur at a certain point. Two simple models from the plasticity theory are commonly used to characterize the cyclic plastic behavior. The *kinematic hardening* model assumes that the yield surface does not change its size or shape but simply translates in stress space in the direction of outward normal; the *isotropic hardening* model, on the other hand, assumes that the yield surface expands uniformly in all directions [6, 7, 12, 13]. In the case of uniaxial loading, reversed yielding predicted by these two models is illustrated schematically in Fig. 2.3. The magnitudes of initial yield strength (during forward loading) and reversed yield strength are σ_y and σ_y^r, respectively. The material is first loaded to a peak stress, σ^{peak}, before unloading and then loading in the opposite direction. According to the kinematic hardening model, the stress span from σ^{peak} to the reversed yielding point is equivalent to the magnitude of $2\sigma_y$. As a consequence reversed yielding occurs when the stress value reaches σ^{peak}-$2\sigma_y$ (the curve labeled "kinematic hardening" in Fig. 2.3). As for the isotropic hardening model, the stress span from σ^{peak} to the reversed yielding point is set to be equal to $2\sigma^{peak}$ (or the value of reversed yield strength is $-\sigma^{peak}$; the curve labeled "isotropic hardening" in Fig. 2.3). Note the two models predict the same cyclic stress–strain response if the material is elastic–perfectly plastic.

The reversed yield strength for actual metallic materials depends on the alloy composition and microstructure. It may be close to either model, or something in between. It is important to acquire experimental information for a given material before choosing the model for carrying out analyses. It is worth mentioning that the kinematic hardening model is able to predict the Bauschinger effect (early reversed

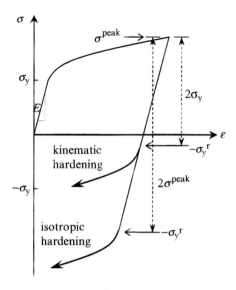

Fig. 2.3 Schematic showing how reversed yielding during uniaxial loading is specified by the models of *kinematic hardening* and *isotropic hardening*. The magnitudes of initial yield strength (during forward loading) and reversed yield strength are denoted by σ_y and σ_y^r, respectively. The symbol σ^{peak} denotes the largest stress at the end of the forward loading

yielding, or $\sigma_y^r < \sigma_y$) observed in many engineering alloys. Furthermore, when considering a true cyclic loading process (i.e., two reversals or more), the kinematic hardening model will predict a stabilized stress–strain hysteresis loop while the isotropic hardening model will lead to continued cyclic hardening until "elastic shakedown" [12].

2.3.2 Multiaxial Response

From the materials science point of view, plastic deformation is controlled by the slip of dislocations along with other possible microscopic mechanisms such as climb of dislocations, mechanical twinning, diffusional flow and relative movements of individual grains etc [9, 10]. The mechanics is complex even with a phenomenological approach. There are several plastic yielding criteria applicable to cases where multiaxial stresses are involved [6, 7]. Here we focus on the von Mises formulation, which is the most representative and commonly employed criterion.

Plastic flow is caused by the deviatoric part of the stress field. The deviatoric stress tensor σ'_{ij} is defined to be

$$\sigma'_{ij} = \sigma_{ij} - \delta_{ij}\sigma_H, \tag{2.14}$$

where σ_{ij} represents the general stress components and δ_{ij} is the Kronecker delta ($\delta_{ij} = 1$ if $i = j$; $\delta_{ij} = 0$ if $i \neq j$). In terms of the matrix representation, the deviatoric stress tensor is $\begin{bmatrix} \sigma_{11} - \sigma_H & \sigma_{12} & \sigma_{13} \\ \sigma_{12} & \sigma_{22} - \sigma_H & \sigma_{23} \\ \sigma_{13} & \sigma_{23} & \sigma_{33} - \sigma_H \end{bmatrix}$. Here the hydrostatic stress tensor $\delta_{ij}\sigma_H$ is subtracted from the general stress tensor σ_{ij}. Note the hydrostatic stress causes only a volume change in an isotropic material, while the deviatoic stress components are responsible for the change in shape. Plastic deformation is volume conservative and is not affected by the hydrostatic stress.

The von Mises effective stress, σ_e, is defined as

$$\sigma_e = \frac{1}{\sqrt{2}}\left[\left(\sigma_{11} - \sigma_{22}\right)^2 + \left(\sigma_{22} - \sigma_{33}\right)^2 + \left(\sigma_{33} - \sigma_{11}\right)^2 + 6\left(\sigma_{12}^{\ 2} + \sigma_{23}^{\ 2} + \sigma_{13}^{\ 2}\right)\right]^{\frac{1}{2}}. \quad (2.15)$$

Plastic yielding commences when the magnitude of σ_e reaches σ_y, the yield strength of the material under uniaxial loading. In the special case of uniaxial loading along the 1-direction, $\sigma_{22} = \sigma_{33} = \sigma_{12} = \sigma_{23} = \sigma_{13} = 0$ and (2.15) is reduced to

$\sigma_e = |\sigma_{11}|$. Note $\sigma_e = \sqrt{3J_2} = \sqrt{\frac{3}{2}\sigma'_{ij}\sigma'_{ij}}$, where J_2 is the second invariant of the deviatoric stress tensor σ'_{ij}. The von Mises yield condition, in conjunction with the associated flow rule [6, 7], is commonly employed in computational modeling of metal deformation involving plasticity.

Upon yielding, the total strain of an elastic-plastic material, ε_{ij}, is the sum of the elastic part ε_{ij}^c and the plastic part ε_{ij}^p. The incremental flow theory relates the increment of plastic deformation to stress in the functional form of

$$\sigma_e = h\left(\int d\bar{\varepsilon}^p\right) \quad (2.16)$$

where h is the strain hardening function and $d\bar{\varepsilon}^p$, the effective plastic strain increment, is

$$d\bar{\varepsilon}^p = \sqrt{\frac{2}{3}d\varepsilon_{ij}^p d\varepsilon_{ij}^p}$$

$$= \frac{\sqrt{2}}{3}\left[\left(d\varepsilon_{11}^p - d\varepsilon_{22}^p\right)^2 + \left(d\varepsilon_{22}^p - d\varepsilon_{33}^p\right)^2 + \left(d\varepsilon_{33}^p - d\varepsilon_{11}^p\right)^2 + 6\left(d\varepsilon_{12}^{p\ 2} + d\varepsilon_{23}^{p\ 2} + d\varepsilon_{13}^{p\ 2}\right)\right]^{\frac{1}{2}}. \quad (2.17)$$

The strain hardening function h in (2.16) can follow (2.13) or other similar forms, or in a numerical analysis it may simply consist of tabulated data that fit the stress–strain curve of an actual material. Note in the special case of uniaxial stressing along the 1-direction, $d\varepsilon_{22}^p = -\frac{1}{2}d\varepsilon_{11}^p$ and $d\varepsilon_{33}^p = -\frac{1}{2}d\varepsilon_{11}^p$ due to volume conservation, and (2.17) is reduced to $d\bar{\varepsilon}^p = d\varepsilon_{11}^p$. The individual incremental plastic strain components are related to the stress components by

$$d\varepsilon_{ij}^p = \frac{3}{2}\frac{\sigma'_{ij}}{\sigma_e}d\bar{\varepsilon}^p. \quad (2.18)$$

After the material has experienced a plastic deformation history, the equivalent plastic strain (or effective plastic strain) is then

$$\bar{\varepsilon}^p = \int_0^t \frac{d\bar{\varepsilon}^p}{dt} \, dt, \tag{2.19}$$

where t is the time history.

It is noted that the treatment above is given in its most accessible form. The yield strength and strain hardening response of the metal need to be known in order to carry out an analysis. A main limitation of continuum plasticity is that, when discrete microscopic phenomena become dominant at very small length scales, quantitative information obtained from the continuum analysis becomes less representative. However, the predicted qualitative trend and spatial deformation pattern will still be useful.

2.4   Time-Dependent Deformation

2.4.1   Viscoplastic Response

When mechanical loading occurs at elevated temperatures, deformation may become a function of time so more refined constitutive models are needed. (Note this is also true for low-melting-point alloys or polymers even at room temperature.) The time-dependent nature can also be manifested in the form that material strength is a function of strain rate (not just strain). Here we present a simple *viscoplastic* model, which is a direct extension of plasticity to incorporate the strain rate effect. It is based on the static stress–strain relation in (2.16), with a scaling parameter to quantify the "strain rate hardening" effect,

$$\sigma_e = h\left(\int d\bar{\varepsilon}^p\right) \cdot R\left(\frac{d\bar{\varepsilon}^p}{dt}\right). \tag{2.20}$$

where h (as a function of plastic strain) is the static plastic stress–strain response, and R, a function of plastic strain rate $\dfrac{d\bar{\varepsilon}^p}{dt}$, defines the ratio of flow stress at non-zero strain rate to the static flow stress (where $R = 1.0$). The functional form of R is given by the experimental data of a specific material. Figure 2.4 shows a schematic of stress–strain curves illustrating the increase in plastic flow stress as the applied strain rate increases. The general shape of the stress–strain curve is used for determining h and the relative vertical positions of the curves are used for determining R. Here an implicit assumption is that the general multiaxial viscoplastic response also follows that of the uniaxial case.

Frequently an empirical expression relating stress and strain rate is employed,

$$\sigma = C\dot{\varepsilon}^m, \tag{2.21}$$

where C is a material constant bearing the rate-independent stress–strain information, $\dot{\varepsilon}$ is the time rate of strain, and m is a material parameter termed strain rate sensitivity. Note (2.21) may be viewed as a special form of the generic expression (2.20).

Another commonly used technique to quantify the time-dependent deformation behavior is the uniaxial constant-stress creep test. Here the change of strain with time is monitored under a constant applied stress. After a transient period, a steady state is reached where the strain rate remains nearly constant. From the physical point of view, the steady-state strain rate can be written in the general functional form:

$$\dot{\varepsilon}_s = f(\sigma, T, S, P),\qquad(2.22)$$

where σ is the applied stress, T is temperature, S represents the state variables describing the microstructural state of the material such as grain size and dislocation density, and P includes material characteristics such as lattice parameter, atomic volume, bond energies, elastic modulus, diffusion constants, etc.[14] Equations specific to the various relaxation mechanisms, including dislocation glide, climb-controlled dislocation glide (power-law creep), and diffusional creep, are available. More convenient for engineering analyses is the following expression,

$$\dot{\varepsilon}_s = A\sigma^{n_c} \exp\left(-\frac{Q}{RT}\right),\qquad(2.23)$$

where A is a constant, σ is the applied stress, n_c is the stress exponent for creep, Q is the activation energy, R is the universal gas constant (8.314 J/mol K), and T is the absolute temperature. The parameters A, n_c and Q for a specific material under given stress and temperature ranges are determined from experimental measurements.

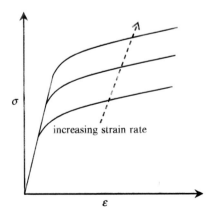

increasing strain rate

Fig. 2.4 Schematic showing the influence of applied strain rate on the plastic flow stress of an elastic-viscoplastic material

2.4.2 Viscoelastic Response

Another type of time-dependent deformation, typically for amorphous polymeric solids around or above the glass transition temperature, is viscoelasticity. Here when a stress is applied, the strain will not stay constant but increases with time. Upon release of the applied load, the strain returns to zero over time. This time-dependent behavior is a manifestation that it incorporates a mixture of both elastic and viscous characteristics: the material is elastic in that it recovers but is viscous in that it creeps. If the ratio of stress and viscoelastic strain is a function of time only, the response is called linear viscoelastic. This normally occurs when the elastic strains and viscous flow rates are small. In addition to creep, a viscoelastic material displays stress relaxation in that when a fixed strain is imposed, the induced stress will decrease over time.

Continuum models describing the linear viscoelastic response commonly involves serial and/or parallel arrangements of linear springs and dashpots. The instantaneous elastic response is represented by the spring element of elastic modulus E, where the uniaxial stress and strain is related by $\sigma = E\varepsilon$. The viscous response is represented by the dashpot of viscosity η, which relates the stress and strain rate through $\sigma = \eta(d\varepsilon / dt)$. Figure 2.5 shows three simple one-dimensional examples of such models under an applied stress σ.

The Maxwell model, Fig. 2.5a, consists of a spring and a dashpot in series. Its governing constitutive relation is

$$\frac{d\varepsilon}{dt} = \frac{\sigma}{\eta} + \frac{1}{E}\frac{d\sigma}{dt}. \qquad (2.24)$$

The Kelvin–Voigt model is also a combination of a spring and a dashpot but in a parallel arrangement, Fig. 2.5b. The governing equation becomes

$$\sigma = \eta\frac{d\varepsilon}{dt} + E\varepsilon. \qquad (2.25)$$

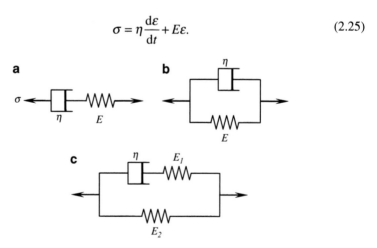

Fig. 2.5 Schematic arrangements of (**a**) the Maxwell model, (**b**) the Kelvin-Voigt model, and (**c**) a three-element standard linear model for linear viscoelastic materials

Figure 2.5c shows a type of so called standard linear solid model, which is composed of a spring element in parallel with a Maxwell model. Applying the basic mechanical relations of the elements and taking into account the series and parallel features involved, one can obtain the constitutive equation for the model in Fig. 2.5 as

$$\frac{d\varepsilon}{dt} = \left(\frac{1}{E_1 + E_2}\right)\left(\frac{d\sigma}{dt} + \frac{E_1}{\eta}\sigma - \frac{E_1 E_2}{\eta}\varepsilon\right). \tag{2.26}$$

It is noted that multi-component springs and dashpots are frequently needed to characterize realistic viscoelastic behavior. For instance, generalized Maxwell models, featuring a number of Maxwell elements in parallel, or generalized Kelvin–Voigt models, featuring a number of Kelvin–Voigt elements in series, are commonly employed to fit the time-dependent deformation with higher degrees of precision. For general 3D loading, the purely hydrostatic state of stress leads to almost perfect elastic deformation and the above equations about creep and viscoelasticity can be re-derived using the deviatoric components of stress and strain [4].

2.5   Thermal Expansion and Thermal Mismatch

The vast majority of materials expand upon heating and contract upon cooling. Therefore a temperature change ΔT induces a strain, $\alpha \cdot \Delta T$, in all three dimensions. Here α is the linear coefficient of thermal expansion (CTE), a material property. For an isotropic elastic-plastic solid, the normal strains under the influence of applied stress (mechanical loading) and temperature change (thermal loading) can thus be written as

$$\varepsilon_{11} = \frac{1}{E}\left[\sigma_{11} - v(\sigma_{22} + \sigma_{33})\right] + \varepsilon_{11}^p + \alpha \cdot \Delta T$$

$$\varepsilon_{22} = \frac{1}{E}\left[\sigma_{22} - v(\sigma_{11} + \sigma_{33})\right] + \varepsilon_{22}^p + \alpha \cdot \Delta T \tag{2.27}$$

$$\varepsilon_{33} = \frac{1}{E}\left[\sigma_{33} - v(\sigma_{11} + \sigma_{22})\right] + \varepsilon_{33}^p + \alpha \cdot \Delta T$$

where ε_{ij}^p represents the accumulated plastic part of the strain component. The temperature change does not affect the shear strain components.

When a material element, free of any applied mechanical stress, is able to change its dimensions *freely* upon heating or cooling, no stress will be generated. If, however, a non-uniform temperature distribution exists in the material (due to, e.g., external thermal conditions and/or transient heat conduction), internal stresses can be induced as a result of the mismatch in dimensional change between adjacent material elements. Even in the absence of non-uniform temperature field, stresses may still develop if the material is under physical confinement. Below we present two simple analytic examples as an illustration.

2.5.1 Example: Residual Stress Buildup in a Constrained Rod

The first example concerns the thermally induced elastic-plastic deformation of a rod rigidly clamped at the two ends, as shown in Fig. 2.6a. For simplicity only the longitudinal stress and strain are considered so the problem becomes one-dimensional in nature. We further assume that the material is elastic-perfectly plastic with all properties (Young's modulus E, CTE α, and yield strength σ_y) independent of temperature. The stress–strain history is shown in Fig. 2.6b. Starting from a stress-free condition, the rod is subject to heating (with positive ΔT). The total strain ε_{tot} is thus the sum of thermal, elastic and plastic strains (ε^t, ε^e, and ε^p, respectively),

$$\varepsilon_{tot} = \varepsilon^t + \varepsilon^e + \varepsilon^p = \alpha \cdot \Delta T + \frac{\sigma}{E} + \varepsilon^p. \tag{2.28a}$$

Upon heating, the rod tends to expand but is forced to remain a constant length so it is under compressive stress. Plastic yielding occurs if the magnitude of $\alpha \cdot \Delta T$ is greater than that of the yield strain, σ_y / E. Setting (2.28a) to zero and σ to $-\sigma_y$ for yielding, one obtains

$$\varepsilon^p = -\alpha \cdot \Delta T + \frac{\sigma_y}{E} < 0. \tag{2.28b}$$

Upon cooling, elastic recovery takes place and the stress continues into the positive regime. When the temperature change is reduced to zero, a tensile residual stress equal to $-E\varepsilon^p$ is left in the material (Fig. 2.6b). It is possible that, if ΔT is sufficiently large, reversed yielding in tension may occur during cooling. This type of thermal residual stress is commonly seen in traditional engineering alloys (after, e.g., welding and heat treatment), as well as in advanced heterogeneous materials and devices under physical constraint.

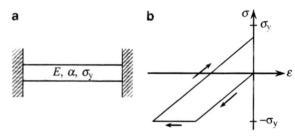

Fig. 2.6 Schematics showing (**a**) the clamped rod and (**b**) the stress-strain history during the heating and cooling process

2.5.2　Example: Thermoelastic Deformation of Bi-material Layers

In the second example we consider a layered structure consisting of two different materials subjected to a uniform temperature change. The thermal expansion mismatch between the two layers induces internal stresses and an overall shape change (curvature) of the initially flat plate. This is the classical problem of bi-metal thermostats, the mechanics of which was systematically investigated by Timoshenko [15]. In the present example we present a general analysis focusing on linear elasticity within the context of small deformation. A universal relation of curvature evolution is derived, and it is presented in graphical form for all possible combinations of layer geometry and elastic modulus [16].

Consider the bi-material structure composed of perfectly bonded layer 1 of thickness h_1 and layer 2 of thickness h_2, as shown in Fig. 2.7a. The dimension in x is much greater than that in z. The material properties of concern are Young's modulus E, Poisson's ratio v, and CTE α of the respective layers. The layers are assumed to be stress-free and flat at an initial temperature ($\Delta T = 0$). We focus on regions away from the remote free ends, and the plane stress assumption ($\sigma_{yy} = 0$) is first employed in the analysis. The variation of longitudinal strain ε_{xx} along the z-axis is given by the beam kinematics,

$$\varepsilon_{xx} = \varepsilon = \varepsilon_0 + \kappa z, \tag{2.29}$$

where ε_0 is the strain at the interface and κ is the curvature. This relation conforms to the pure bending assumption that any cross section perpendicular to the axis of the beam remains planar and perpendicular to the curved axis during bending. The stresses in the two layers are

$$\sigma_1 = \sigma_{xx.1} = E_1 \left(\varepsilon - \alpha_1 \Delta T \right) \quad \text{and} \quad \sigma_2 = \sigma_{xx.2} = E_2 \left(\varepsilon - \alpha_2 \Delta T \right). \tag{2.30}$$

Equilibrium of forces and moments gives

$$\int_{-h_2}^{0} \sigma_2 dz + \int_{0}^{h_1} \sigma_1 dz = 0 \quad \text{and} \quad \int_{-h_2}^{0} \sigma_2 z dz + \int_{0}^{h_1} \sigma_1 z dz = 0, \tag{2.31}$$

respectively. Combining (2.29–2.31) and solving for ε_0 and κ, one obtains

$$\varepsilon_0 = \frac{\Delta T}{\gamma} \left\{ E_1^2 h_1^4 \alpha_1 + E_2^2 h_2^4 \alpha_2 + E_1 E_2 h_1 h_2 \left[4h_2^2 \alpha_1 + 4h_1^2 \alpha_2 + 3h_1 h_2 \left(\alpha_1 + \alpha_2 \right) \right] \right\} \tag{2.32}$$

and

$$\kappa = \frac{\beta}{\gamma} \cdot 6 h_1 h_2 \left(h_1 + h_2 \right), \tag{2.33}$$

where

$$\beta = E_1 E_2 \left(\alpha_1 - \alpha_2 \right) \Delta T \tag{2.34}$$

And

$$\gamma = E_1^2 h_1^4 + E_2^2 h_2^4 + E_1 E_2 h_1 h_2 \left(4h_1^2 + 6h_1 h_2 + 4h_2^2 \right). \tag{2.35}$$

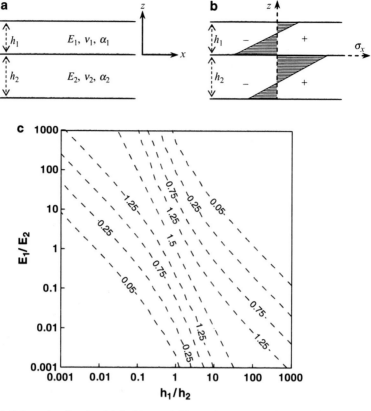

Fig. 2.7 Schematics showing (**a**) the bi-material layered structure and the coordinate system, and (**b**) the qualitative stress profile within each layer induced by thermal expansion mismatch. (**c**) Contours of the normalized elastic curvature $\bar{\kappa}$ in the domain defined by the elastic modulus ratio E_1 / E_2 and the layer thickness ratio h_1 / h_2

The stress field within each layer is obtained by substituting (2.29), (2.32) and (2.33) back to (2.30). The stress varies linearly within each layer, as schematically depicted in Fig. 2.7b, and the values at the free surfaces and interface are

$$\sigma_{1,z=h_1} = \frac{\beta}{\gamma}\{E_1 h_1 h_2 (2h_1^2 + 3h_1 h_2) - E_2 h_2^4\}$$

$$\sigma_{1,z=0} = -\frac{\beta}{\gamma}\{E_2 h_2^4 + E_1 h_1 h_2 (4h_1^2 + 3h_1 h_2)\}$$

$$\sigma_{2,z=0} = \frac{\beta}{\gamma}\{E_1 h_1^4 - E_2 h_1 h_2 (2h_2^2 + 3h_1 h_2)\}$$

$$\sigma_{2,z=-h_2} = \frac{\beta}{\gamma}\{E_1 h_1^4 + E_2 h_1 h_2 (4h_2^2 + 3h_1 h_2)\}.$$

(2.36)

Note the greatest magnitude of stresses within each layer appears at the interface. This maximum stress at the interface is compressive in layer 1 and is tensile in layer 2 for the case $\alpha_1 > \alpha_2$ and $\Delta T > 0$.

The overall configuration change of the bilayer beam due to uniform heating or cooling is given by (2.33), which can be rearranged as

$$\kappa\left(h_1 + h_2\right) = \left(\alpha_1 - \alpha_2\right)\Delta T \cdot \frac{6\dfrac{h_1}{h_2} + 6\dfrac{h_2}{h_1} + 12}{\dfrac{E_1}{E_2}\left(\dfrac{h_1}{h_2}\right)^2 + \dfrac{E_2}{E_1}\left(\dfrac{h_2}{h_1}\right)^2 + 4\dfrac{h_1}{h_2} + 4\dfrac{h_2}{h_1} + 6} \qquad (2.37)$$

It is noticed that the value $\kappa\left(h_1 + h_2\right)$ is proportional to the difference of thermal strain, $\left(\alpha_1 - \alpha_2\right)\Delta T$, when the modulus ratio and thickness ratio of the two layers are fixed. This suggests the use of a normalized curvature,

$$\bar{\kappa} = \frac{\kappa\left(h_1 + h_2\right)}{\left(\alpha_1 - \alpha_2\right)\Delta T} = \frac{6\dfrac{h_1}{h_2} + 6\dfrac{h_2}{h_1} + 12}{\dfrac{E_1}{E_2}\left(\dfrac{h_1}{h_2}\right)^2 + \dfrac{E_2}{E_1}\left(\dfrac{h_2}{h_1}\right)^2 + 4\dfrac{h_1}{h_2} + 4\dfrac{h_2}{h_1} + 6}, \qquad (2.38)$$

which is a dimensionless quantity. Figure 2.7c shows a contour plot of constant $\bar{\kappa}$ in a domain defined by E_1 / E_2 and h_1 / h_2. The range of thickness ratio chosen is such that the two extremes of thin-film/thick-substrate combinations for most engineering applications are included. It is seen that a line of maximum value of $\bar{\kappa}$ exists. This can in fact be shown by differentiating $\bar{\kappa}$ in (2.38). For example, the expression

$$\frac{d\bar{\kappa}}{d\left(\dfrac{E_1}{E_2}\right)} = 0 \qquad (2.39)$$

can be worked out and simplified to

$$\frac{E_1}{E_2} = \left(\frac{h_1}{h_2}\right)^{-2}, \qquad (2.40)$$

which represents the line of maximum $\bar{\kappa}$ in Fig. 2.7. The maximum value can be obtained by substituting (2.40) back to (2.38), which results in a value of 3/2. Therefore, 3/2 is the greatest normalized curvature one can obtain in all combinations of materials and their thicknesses, as seen in Fig. 2.7. It can be seen in this diagram that when one layer is significantly thicker than the other, the curvature is very small unless the thin layer is extremely stiffer than the thick layer.

In all the derivations above we assumed a plane stress condition, with the dimension of the beam in the y-direction very small. In a more general situation involving a bi-material plate rather than beam, the stress state will be equi-biaxial ($\sigma_{xx} = \sigma_{yy}$) and all results in this section still hold true with the only exception that the Young's modulus E for each layer is replaced by the corresponding biaxial modulus

$E/(1-v)$ where v is the Poisson's ratio. When a plane strain state of deformation exists, the plane strain modulus $E/(1-v^2)$ should be used instead and α needs to be replaced by $(1+v)\alpha$. Note in an actual plate-like layered structure, bifurcation in equilibrium shape may occur if the mismatch effect is large enough to trigger geometrically nonlinear deformation. A different approach is needed under this circumstance [17].

2.6 The Numerical Modeling Approach

In many device structures, the geometries and/or the material properties are too complex to be dealt with by simple analytical means such as in the examples in Sects. 2.5.1 and 2.5.2. Numerical techniques are therefore sought for the deformation analysis. The finite element method implemented on a computer is the most widely used technique for this purpose. The modeled structure is divided into small elements connected to each other at nodal points. Generally (and superficially) speaking, the method involves constructing a matrix equation of the form

$$\{P\} = [K]\{u\}, \tag{2.41}$$

where $\{P\}$ represents the loads and reactions at the nodal points, $[K]$ represents the stiffness matrix assembled based on the mechanical properties of the material and the geometry of the finite elements, and $\{u\}$ represents the displacements at the nodes to be solved by the analysis. Once the displacements are solved, the strain and stress fields can then be computed through the strain–displacement and stress–strain relations, respectively. The reader may refer to specialized books on the finite element method for theoretical background and numerical implementation techniques [18–20].

2.6.1 Example: Sanity Test

There have been many commercial and in-house finite element computational programs developed for a wide variety of applications. While learning to use the computational tools may be relatively straightforward, it is of particular importance that the user understands the definition of parameters and fundamental mechanics involved before embarking on an analysis. Here we present a seemingly trivial example as an illustration. The problem is to apply the finite element method to model uniaxial loading of a homogeneous elastic-plastic material. The task at hand is to ensure the stress–strain response obtained as output from the finite element analysis is equivalent to that given as input. Figure 2.8 shows a schematic of a cylindrical specimen subject to axial stretching along the z-direction. This 3D scheme can be simplified to a 2D one by invoking the axisymmetric type of elements.

The rectangular domain needed for the analysis is highlighted in Fig. 2.8a and shown in Fig. 2.8b along with the imposed boundary conditions. The left boundary is the axis of symmetry and no displacement in the r-direction is allowed. The bottom boundary can be viewed as the mid cross section of the cylinder so displacement in the z-direction is forbidden. The applied displacement is imposed on the top boundary and the reaction force is generated. The right boundary is the free surface.

The finite element program Abaqus (Dassault Systemes Simulia Corp., Providence, RI) is employed in this example and many other case studies in this book. The Young's modulus and Poisson's ratio of the material are taken to be 70 GPa and 0.33, respectively. The entire true stress–true strain relation used as input is shown in Fig. 2.8c (the curved labeled as "input"). Note the plastic portion is made to be

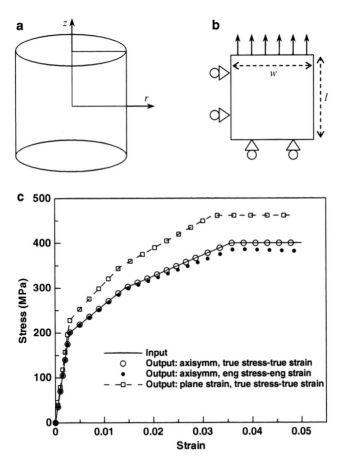

Fig. 2.8 (**a**) Schematic of a cylindrical specimen and the 2D rectangular region used for the finite element modeling. (**b**) The computational domain along with the boundary conditions. The displacement is imposed on the top boundary to simulate uniaxial stretching. (**c**) Comparison of the modeled stress–strain curves with the stress–strain behavior used as input in the finite element analysis

piecewise linear with an initial yield strength of 200 MPa, and the flow strengths at the plastic strain values of 0.01 and 0.03 are taken to be 300 and 400 MPa, respectively. Beyond 400 MPa the material follows a perfectly plastic behavior. The output stress–strain curves, shown in Fig. 2.8c, are obtained from the load–displacement data of the finite element analysis. Two output responses from the axisymmetric model are considered: one using the true stress and true strain definition and the other using the engineering stress and engineering strain definition. Here the true stress is calculated by dividing the load by the *current* cross section area ($w^2\pi$) obtained from the modeling, and the engineering stress is calculated by dividing the load by the original cross section area before deformation. The true strain and engineering strain are calculated according to, respectively, $\ln(l/l_0)$ and $(l-l_0)/l_0$, where l and l_0 represent the current and original length of the computational domain, respectively. It can be seen from Fig. 2.8c that the true stress–true strain output from the axisymmetric model matches the input curve throughout, suggesting that the numerical model was set up correctly. The engineering stress–engineering strain output, on the other hand, cannot reproduce the true input.

For comparison purposes an additional case of output is included in Fig. 2.8c: the plane strain state applied to the model in Fig. 2.8b with no displacement allowed in the out-of-paper direction. It can be seen that the plane strain assumption produces a result far from close to the actual material response. This is due to the suppression of contraction in the out-of-paper direction so a higher longitudinal stress needs to be applied to achieve the same extent of stretching. The slope of the elastic portion is equivalent to the plane strain modulus $E/(1-v^2)$. As for the plastic part, it can be shown from the von Mises yield criterion that the flow stress corresponding to the plane strain model differs from the actual uniaxial value by a factor of $2/\sqrt{3}$.

The example above, while straightforward, illustrated the importance that one has to "know what he (she) is doing" to avoid wasteful efforts and ensure the generation of meaningful results using the numerical modeling approach. When using commercial programs (or any computer code for that matter), one is also advised to pay attention to the possible differences in their "default" settings such as the definition of stress and strain.

2.6.2   Example: Proper Interpretation and Use of the Output

After the output is generated from the modeling, it is also essential to interpret the result with the right mechanistic understanding in mind. In this section we give another simple example that deals with the identification of internal stress field in a loaded structure. The specimen considered is a thin square shaped frame-like object, as shown in Fig. 2.9a, under the plane stress condition. The boundary near the lower-left corner is fixed in space while the upper-right boundary is subject to pulling along the 45° direction. The extent of deformation is such that the horizontal and vertical components of the pulling displacement are 2% of the initial outer

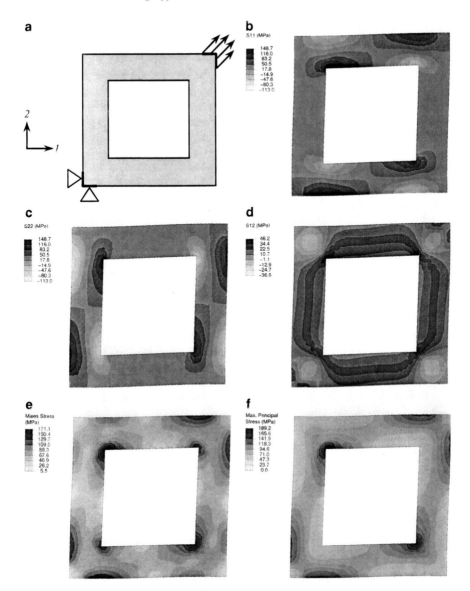

Fig. 2.9 (**a**) Schematic of the square shaped frame-like structure along with the loading and boundary conditions used in the finite element modeling. (**b–f**) The deformed configuration and contour plots of (**b**) σ_{11} normal stress, (**c**) σ_{22} normal stress, (**d**) σ_{12} shear stress, (**e**) von Mises effective stress, and (**f**) maximum principal stress

side length of the structure. The material is purely elastic with Young's modulus 10 GPa and Poisson's ratio 0.3. The choices of material, geometry and loading for this example are arbitrary. The modeling objective is to ensure the largest tensile stress is within a certain limit so as to prevent, for instance, brittle fracture under the prescribed deformation.

Figures 2.9b to f show the contour plots of the stress components σ_{11}, σ_{22}, σ_{12}, von Mises effective stress and maximum principal stress, respectively. The distorted shape of the structure is evident. It can be seen that the patterns of σ_{11} and σ_{22} form a mutual mirror reflection about the 45° axis, as expected due to the symmetric setup of the problem. There is also a significant shear stress field of σ_{12}. The maximum value of σ_{11} (and σ_{22}) is 148.7 MPa. Note this is not the maximum tensile stress in the material because the stress values of the individual components are a consequence of the chosen coordinate system. When one considers the von Mises effective stress field, Fig. 2.9e, a maximum value of 171.1 MPa is seen. Note this is still not the largest stress we are seeking because the von Mises stress is defined to be an all-positive scalar quantity dominated by the deviatoric contribution of the stress tensor. It may be used as an indication of the propensity of plastic yielding, but not for brittle materials where crack initiation is typically a result of local tensile stress. In many finite element analysis programs the von Mises effective stress is set to be the "default" option of the output stress field in their postprocessors. Unfortunately this has frequently prompted modelers to indiscriminately use it to quantify the highest stress and identify the "danger zone" in their material, even if the concern is brittle fracture. As can be seen in Fig. 2.9f, the highest maximum principal stress appears at the upper-left and lower-right inner corners and attains a value of 189.2 MPa, which is greater than the greatest of σ_{11}, σ_{22} and von Mises stress. It is this highest value of maximum principal stress that should be used to compare against the design limit.

The analysis above illustrated the importance of utilizing the right output parameter to correlate with the physical attribute of concern. In subsequent chapters we will attend to more real-life examples of similar nature that are of direct relevance to constrained deformation in small devices and structures.

2.7 Choosing the Appropriate Constitutive Model

Linear elasticity is the most commonly used constitutive material model in thermomechanical modeling. This is largely due to its simplicity – all it needs is to define two elastic constants for an isotropic material, and the linear analysis normally leads to high computational efficiency. Indeed when the deformation is sufficiently small, all classes of materials including metals, ceramics and polymers can be described by the linear elastic behavior reasonably well. Note that the use linear elasticity is not only common in numerical modeling, it is also the basis in the majority of experimental analyses involving, for instance, the conversion of stress and strain.

In reality, all materials display certain degrees of nonlinearity and time dependency at all temperatures under all possible loading rates. The issue then becomes one of choosing a reasonable material model which is appropriate for the given application. In general, ceramic materials are represented well by linear elasticity because there is little or no plasticity involved during deformation until failure, unless the temperature is very high. For metallic materials, an elastic-plastic model is the most suitable. When the yield condition is met, the movement of dislocations inside metallic crystals leads to permanent deformation and possible work hardening, which are depicted by plasticity theory. If deformation occurs at temperatures greater than about $0.4T_m$ (where T_m is the absolute melting point of the material), then time-dependent plastic deformation may need to be considered. As for typical polymeric materials, a linear elastic behavior is generally sufficient at below the glass transition temperature (T_g). Above T_g the viscoelastic effect may become significant. Note in viscoelastic creep, the molecular chain movement is such that the deformation is recoverable upon the removal of applied load. In some polymers with linear chain configurations, however, the inter-chain slippage may lead to permanent deformation so plastic yielding will need to be accounted for.

To this end it is important to note that the material properties themselves (elastic modulus, yield strength etc.) are functions of temperature. If the problem at hand involves a large temperature change, this temperature dependence may need to be incorporated in the analysis for increased accuracy.

Although the treatment in this chapter is limited to isotropic materials, it is also important to recognize that isotropy is only an approximation under certain circumstances. If, for example, the structural dimension is essentially that of a single crystal and the material property possesses a high degree of directionality, then the anisotropic constitutive relation may need to be invoked. Special geometric arrangements of the micro-constituents can also lead to highly anisotropic effective properties of materials. There will be further discussion on this in subsequent chapters when the relevant cases are presented. For detailed theoretical background on the thermo-mechanical behavior involving anisotropy, the reader is referred to other useful sources [7, 21, 22].

References

1. G. T. Mase and G. E. Mase (1999) Continuum mechanics for engineers, 2nd ed., CRC Press, Boca Raton, Florida.
2. L. E. Malvern (1969) Introduction to the mechanics of a continuous medium, Prentice-Hall, Englewood Cliffs, New Jersey.
3. Y. C. Fung (1965) Foundations of solid mechanics, Prentice-Hall, Englewood Cliffs, New Jersey.
4. B. A. Boley and J. H. Weiner (1997) Theory of thermal stresses, Dover Publications, Mineola, New York.
5. S. P. Timoshenko and J. N. Goodier (1970) Theory of elasticity, McGraw-Hill, New York.
6. A. Mendelson (1968) Plasticity: theory and application, MacMillan, New York.
7. R. Hill (1950) The mathematical theory of plasticity, Oxford University Press, Oxford.
8. M. F. Kanninen and C. H. Popelar (1985) Advanced fracture mechanics, Oxford University Press, New York.
9. J. Roesler, H. Harders and M. Baeker (2007) Mechanical behavior of engineering materials, Springer, New York.

10. W. Soboyejo (2003) Mechanical properties of engineered materials, Marcel Dekker, New York.
11. W. Ramberg and W. R. Osgood (1943) "Description of stress–strain curves by three parameters," Technical Note No. 902, National Advisory Committee for Aeronautics, Washington D.C.
12. S. Suresh (1998) Fatigue of materials, 2nd ed., Cambridge University Press, Cambridge.
13. T. J. Chung (2007) General continuum mechanics, Cambridge University Press, Cambridge.
14. H. J. Frost and M. F. Ashby (1982) Deformation mechanism maps, Pergamon Press, Oxford.
15. S. Timoshenko (1925) "Analysis of bi-metal thermostats," Journal of Optical Society of America, vol. 11, pp. 233–255.
16. C. T. Lin (1996) "Thermally induced deformation of multi-layered materials: analytical and engineering formulations," Master Thesis, Massachusetts Institute of Technology, Cambridge.
17. L. B. Freund and S. Suresh (2003) Thin film materials – Stress, defect formation and surface evolution, Cambridge University Press, Cambridge.
18. J. Fish and T. Belytschko (2007) A first course in finite elements, Wiley, New York.
19. O. C. Zienkiewicz, R. L. Taylor and J. Z. Zhu (2005) The finite element method: its basis and fundamentals, 6th ed., Elsevier Butterworth-Heinemann, Oxford.
20. J. N. Reddy (1993) An introduction to the finite element method, 2nd ed., McGraw-Hill, New York.
21. J. F. Nye (1972) Physical properties of crystals – Their representation by tensors and matrices, Oxford University Press, London.
22. W. A. Backofen (1972) Deformation processing, Addison-Wesley, Reading, Massachusetts.

Chapter 3
Thin Continuous Films

One of the simplest types of physical confinement is a thin continuous film (or blanket film) attached to a thick substrate material. Frequently this is a beginning form for creating micro- and nano-scale systems such as the semiconductor devices. Here the focus is on the constraint imposed by the substrate on the thin film. A numerical example is given below as an introductory illustration.

We consider a disk-shaped aluminum (Al) thin film bonded to a silicon (Si) substrate, as schematically shown in Fig. 3.1a. The diameter-to-thickness ratio of the Al film is 12, and the Si substrate is 100 times thicker than Al. The initial condition is such that there is no deformation and internal stress. This type of problem can be dealt with by adopting a 2D axisymmetric model as shown in Fig. 3.1b where the boundary conditions are also indicated. The left boundary (z-axis) is the axial symmetry line. The assembly is then subject to thermal cooling of 300°C without any applied mechanical load. Due to a mismatch in coefficients of thermal expansion (CTE, $23 \times 10^{-6} \, \mathrm{K}^{-1}$ for Al and $3 \times 10^{-6} \, \mathrm{K}^{-1}$ for Si, assuming no temperature dependency), tensile stresses are generated in the Al film. Figure 3.1c shows a contour plot of stress σ_{rr} (normal stress in the radial direction) after cooling, assuming the Al film behaves only elastically. In the figure the left boundary is the axis of symmetry and only the Al film and the top portion of Si are displayed. It can be seen that in-plane tensile stresses exist in Al. The stress field approaches an equibiaxial state in the inner part of the film (close to the axis of symmetry). The complex stress state near the outer edge of the film extends over a range on the order of the film thickness (in accord with the Saint-Venant's principle [1, 2]). If plastic yielding of the metal is incorporated in the model, the stress field will be altered. Figure 3.1d shows such a case, with the yield strength of the elastic-perfectly plastic Al set to be 200 MPa. The plastic response follows the description in Sect. 2.3.2, except there is no strain hardening considered for the film. Comparing Fig. 3.1d with c, a reduction of stress level due to yielding can be seen. Near the center of the film the σ_{rr} field is constant with a tensile stress value of 200 MPa. Along with the essentially zero shear stress and out-of-plane normal stress, the von Mises stress obtained from (2.15) is equal to the yield strength of the material (200 MPa). Figure 3.1e shows the contour plot of equivalent plastic strain. The entire Al film

Y.-L. Shen, *Constrained Deformation of Materials: Devices, Heterogeneous Structures and Thermo-Mechanical Modeling*, DOI 10.1007/978-1-4419-6312-3_3,
© Springer Science+Business Media, LLC 2010

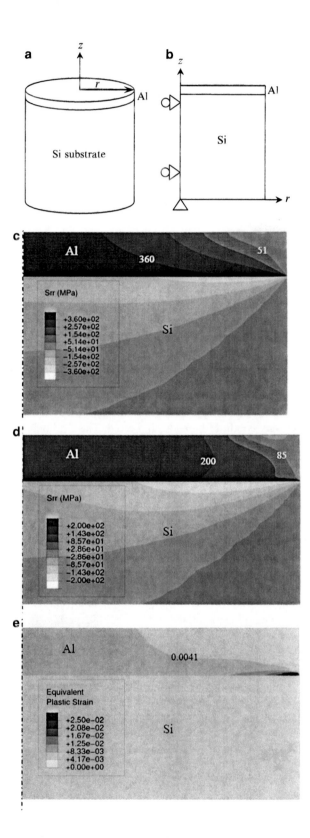

has yielded, but the most severe deformation takes place at the edge of the film adjacent to the substrate. Away from the film edge, the deformation is uniform.

The example in Fig. 3.1 illustrated the following salient features:

1. Significant stresses in the thin film can be generated due to thermal expansion mismatch *alone*, with no external mechanical loading. This is especially common in miniaturized device structures, where materials with different properties are joined together and the structure is frequently subject to temperature variation during processing and service.
2. Plastic yielding in an elastic-plastic metal film can readily occur during thermal loading. Using the simple elastic assumption in analyses may lead to questionable outcomes under many circumstances.
3. If the film is free-standing (not attached to a substrate), the deformation field will be uniform throughout. Bonding to a substrate gives rise to constrained plastic flow. The non-uniformity of deformation near the film edge is particularly distinct. It is anticipated that, if a small-sized material does not have an extensive straight or flat interface with its adjacent material, the edge effect then becomes dominant, resulting in even more complicated deformation patterns.

In the remainder of this chapter, we consider only thin film/substrate systems that have the in-plane dimensions (parallel to the film plane) much greater than the out-of-plane dimension used in the above example, so the film edge effect can be excluded from the discussion. The edge-dominant features will be treated extensively in subsequent chapters.

3.1 Basic Elastic-Plastic Response

3.1.1 Mechanical Loading

When a metal film is bonded to a stiff substrate and the structure is subject to mechanical stretching along an in-plane direction, the strain in the film along the loading direction is equal to that in the substrate and is governed by the applied

---

Fig. 3.1 (a) Schematic showing the axisymmetric model of an Al thin film attached to a Si substrate. (b) The 2D model and boundary conditions used in the simulation. (c) Contours of stress σ_{rr} in Al and the upper portion of Si when Al is treated as purely elastic. (d) Contours of stress σ_{rr} in Al and the upper portion of Si when plastic yielding is allowed in Al. (e) Contours of equivalent plastic strain when plastic yielding is allowed in Al. In (c)–(e) the left boundary represents the symmetry axis, and the thermal history considered is cooling of 300°C from an initial stress-free temperature. The material properties used are: $E_{Al} = 70$ GPa, $v_{Al} = 0.33$, $\sigma_{y, Al} = 200$ MPa, $\alpha_{Al} = 23 \times 10^{-6}$ K^{-1}, $E_{Si} = 130$ GPa, $v_{Si} = 0.28$ and $\alpha_{Si} = 3.0 \times 10^{-6}$ K^{-1}, where E, v, σ_y and α represent Young's modulus, Poisson's ratio, yield strength and coefficient of thermal expansion, respectively

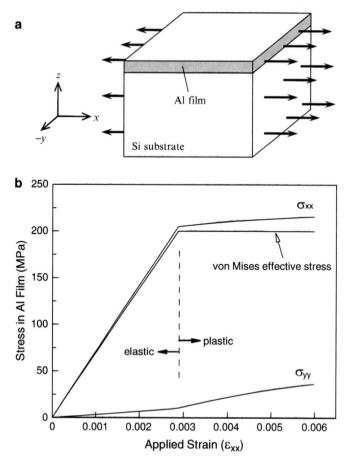

Fig. 3.2 (a) Schematic showing an Al film attached to a Si substrate. The entire structure is subject to uniform stretching (with prescribed boundary displacements). (b) The evolution of stresses σ_{xx}, σ_{yy}, and the von Mises effective stress inside the Al film (The stress σ_{zz} is zero throughout the process)

strain. Figure 3.2a shows a schematic of an Al film on a Si substrate under consideration. The material properties are taken to be the same as in the example in Fig. 3.1. In the present discussion the edge effect mentioned above, i.e., the singular behavior at the lateral edges of the thin film spanning a distance on the order of the film thickness, is ignored. The metal is expected to undergo elastic deformation and later plastic yielding. From intuition one may conceive that a "uniaxial" deformation field exists in the Al film. The deformation, however, is not one-dimensional because the tendency of lateral contraction of the film and the substrate will not be the same. Lateral stresses will thus arise. In this case study the thickness of Si is taken to be five times that of Al. Since the edge effect is excluded, the simulation utilizes the *generalized* plane strain formulation, which is two-dimensional (2D) but accounts for a uniform deformation in the *y*-direction.

Figure 3.2b shows the evolution of stresses σ_{xx}, σ_{yy} and the von Mises effective stress σ_e in Al as a function of applied strain ε_{xx} (σ_{zz} is zero throughout the deformation). The Al film and Si substrate experience the same applied strain ε_{xx}. Due to the greater Poisson's ratio of Al compared to Si (0.33 versus 0.28), tensile stress σ_{yy} starts to build up right from the elastic stage. Upon yielding, a constant von Mises effective stress ensues, the magnitude of which is equal to the Al yield strength (200 MPa). It is important to note that the stress components σ_{xx} and σ_{yy} continue to increase after the onset of yielding for this perfectly plastic material. The lateral constraint imposed by the substrate on the plastically deforming metal is now greater because of volume conservation (with an effective "Poisson's ratio" of 0.5). The increasing σ_{yy} in turn necessitates an increase in σ_{xx} in order to keep up with the applied strain. Although the stress magnitudes σ_{xx} and σ_{yy} are changing, a constant von Mises stress is maintained during plastic deformation as given by (2.15).

This case study illustrated two important features for a substrate constrained metal film:

1. A nominally uniaxial loading can lead to a multiaxial stress state.
2. The individual stress components can continue to increase after yielding, even in a non-strain hardening metal. In other words, in experimental measurements the *apparent* strain hardening given by the measured *axial* stress in metal may be misleading or prone to quantitative errors.

These attributes should be duly accounted for when performing analyses or interpreting experimental data.

The same type of lateral constraint also exists, in a local manner, in deflection tests of microbeams consisting of a metal film on a substrate [3, 4] (see also sections below). While an axial normal stress (or stress amplitude during cyclic tests [5]) in the thin film is generated by the bending process, the film is in fact undergoing biaxial stressing. This stress state may have implications in damage initiation during, for instance, the fatigue test, and thus should not be ignored.

3.1.2 Equi-biaxial Stress State and Thermal Loading

In the case of equi-biaxial deformation, $\sigma_{xx} = \sigma_{yy} = \sigma$ and $\varepsilon_{xx} = \varepsilon_{yy} = \varepsilon$. The elastic stress–strain relation follows

$$\sigma = \frac{E}{1-v} \cdot \varepsilon, \tag{3.1}$$

where $\dfrac{E}{1-v}$ is the biaxial modulus of the material. The von Mises effective stress σ_e becomes simply σ, which is obtained from (2.15) by recognizing that two of the principal stresses are σ and the third vanishes. As the applied stress σ reaches the yield strength of the material σ_y, plastic deformation commences. The most commonly encountered equi-biaxial deformation in a continuous thin film is due to its thermal expansion mismatch with the substrate material, as in the case of Fig. 3.1 but in regions away from the free edge. If the substrate is much thicker than the thin film (by, for instance, at least two orders of magnitude), the biaxial stress in the film can be taken as uniform and the stress value in the elastic state is given by

$$\sigma_f = \frac{E_f}{1-v_f} \cdot \left(\alpha_s - \alpha_f\right) \Delta T, \tag{3.2}$$

where α is the CTE, ΔT is the change in temperature, and the subscripts f and s refer to film and substrate, respectively. If $\alpha_f > \alpha_s$, a compressive stress starts to develop in the metal film during heating ($\Delta T > 0$) and vice versa. With a sufficiently large magnitude of ΔT, σ_f will attain the yield strength σ_y and plastic deformation occurs. If the metal film can strain-harden, the magnitude of σ_f will continue to rise with a further temperature change. For an elastic-perfectly plastic film, however, σ_f stays constant and the metal is essentially free flowing (under equi-biaxial stress σ_y) so as to accommodate any further deformation induced by thermal mismatch.

The stress carried by the metal film can be related to the overall curvature of the film/substrate assembly through the classical Stoney's equation [6],

$$\sigma_f = \frac{E_s}{6(1-v_s)} \frac{h_s^2}{h_f} \cdot \kappa \tag{3.3}$$

where h represents the thickness and κ is the curvature. Note (3.3) only involves the material properties of the substrate material (E_s and v_s), and can be utilized for experimentally determining the film stress during various processes by measuring the curvature [7–14]. If the metal film is capped (passivated) by another thin film, the same methodology can still be applied except that a separate measurement on the passivation/substrate specimen needs to be performed. Because of the large substrate thickness compared to that of the metal and passivation films, each film may be assumed to interact with the substrate independently. The stress in metal can then be extracted [15].

3.1.2.1 Numerical Example

Here we provide a numerical example on the thermal loading induced stress and curvature involving an Al thin film on a thick Si substrate. The problem may be viewed as a trivial exercise, with the purpose of demonstrating a simple but useful modeling technique and correlating the result with the analytic expressions, (3.2) and (3.3). The axisymmetric finite element model utilizes a thin segment of Al and Si as shown schematically in Fig. 3.3a, with the z-axis being the symmetry line along which no displacement in the r-direction is allowed. The right-hand boundary is allowed to move

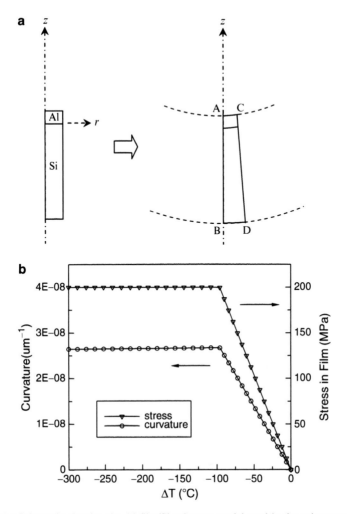

Fig. 3.3 (**a**) Schematic showing the Al film/Si substrate model used in the axisymmetric finite element analysis, before and after the imposed temperature change (drawing not to scale). Only a thin segment of the structure is needed for the analysis if the appropriate boundary conditions are applied. (**b**) Evolution of the σ_{rr} stress in the Al film and curvature κ as a function of the temperature change ΔT

but is constrained to remain straight (linear) in the analysis during deformation (seen as line CD in Fig. 3.3a). This treatment fulfills the "plane-remains-plane" scenario in pure bending, and the film-edge effect is thus excluded. The thicknesses of Al and Si are taken to be 1 and 500 μm, respectively, and their properties are identical to those in the examples in Figs. 3.1 and 3.2. A temperature change of $\Delta T = -300\degree C$ is simulated. During the process the stress in the Al film can be directly obtained from the model output, and the curvature can be simply calculated from the geometric relation between lines AB and CD.

Figure 3.3b shows the evolution of σ_{rr} stress (which is the biaxial stress in this case) and curvature κ as a function of the temperature change ΔT resulting from the modeling. Because the film is very thin compared to the substrate, the stress in Al is thus uniform without any gradient. Elastic deformation in Al occurs up to the temperature change of about -96°C, at which point plastic deformation commences. It can be checked from the figure that the elastic stress–temperature line is in agreement with (3.2). Upon yielding, the biaxial stress stays at a value of 200 MPa in accord with the yield condition. Figure 3.3b also shows that there is a constant ratio between the stress and curvature values at any fixed ΔT, and that (3.3) is followed. Therefore a connection between the numerical modeling and analytical solution is now facilitated. It is interesting to notice that the curvature value after the onset of yielding in Al is, in fact, not strictly constant in Fig. 3.3. It continues to change very slightly with temperature due to the diminutive changes in material thicknesses h_s and h_f in response to the thermal expansion and mismatch effect.

There have been a large number of studies on the strength and inelastic behavior of substrate-bonded thin films, based on the in situ curvature measurement (or similar bending-beam techniques) during temperature cycling [9, 15–51]. The X-ray diffraction technique can also be used for the purpose [29, 33, 52–60]. A form of the stress evolution during temperature cycles is now illustrated. Figure 3.4 shows a representative stress–temperature response of a pure polycrystalline Al film bonded to a thick Si substrate, measured by the wafer curvature method. The film thickness is 1 μm. It shows a hysteresis cyclic loop typically observed after annealing of an as-deposited film at elevated temperature, during which the microstructure in the film becomes stabilized. The initial slopes during heating and cooling represent the thermoelastic response of the Al film. The magnitude of the slope is $\dfrac{E_{Al}}{1-\nu_{Al}}\cdot(\alpha_{Si}-\alpha_{Al})$ as given by (3.2). For example, if α_{Si} and α_{Al} are taken to be 3.0×10^{-6} K^{-1} and 23.0×10^{-6} K^{-1}, respectively, during heating (near room temperature), the initial heating portion of the curve in Fig. 3.4 then gives the *biaxial modulus* of Al film as 111.6 GPa. On the other hand, if α_{Si} and α_{Al} are taken to be 4.9×10^{-6} K^{-1} and 30.0×10^{-6} K^{-1}, respectively, during cooling (near 450°C), the biaxial modulus of Al becomes about 95.6 GPa. The moderate decreasing trend of modulus with increasing temperature can thus be realized.

The deviation of the stress–temperature response from initial linearity signifies plastic yielding. The two dashed lines in Fig. 3.4 are used for fitting the major plastic portion of the film response. The upper and lower lines represent the temperature-dependent tensile yield strength and compressive yield strength, respectively, of the pure Al film. It can be seen that, at a given temperature, the magnitudes of tensile and compressive yield strength are nearly identical. Therefore, the constitutive response of Al films, although thickness and grain size dependent, can be taken as elastic-perfectly plastic with temperature-dependent yield strength. The particular Al film considered in Fig. 3.4 thus shows a room-temperature yield strength magnitude of about 200 MPa. It decreases linearly to about 67 MPa at 400°C.

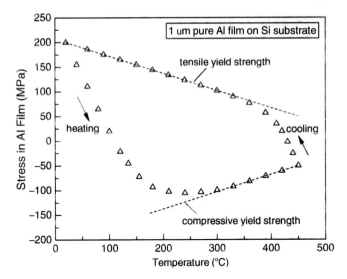

Fig. 3.4 A representative stress-temperature response of a 1-μm thick pure Al film bonded to a Si substrate obtained from the wafer curvature method [30]. A stabilized loop is shown. The two dashed lines are for fitting the plastic flow stress (yield strength) of Al during heating and cooling

3.2 Yielding and Strain Hardening Characteristics

Not all thin metal films display the simple elastic-perfectly plastic response with temperature-dependent yield strength as demonstrated in Fig. 3.4. One prominent example is copper (Cu). When an *unpassivated* Cu is bonded to a substrate (no cap layer above the Cu film), the thermal mismatch induced stress tends to relax significantly at elevated temperature [24, 26, 29, 31, 59]. More discussion on this relaxation behavior in unpassivated Cu films will be given in Sect. 3.6. Passivated (capped) Cu films, however, behave differently and their stress-temperature response is subject to the rate-independent elastic-plastic analysis. This has been attributed to the prevention of surface atomic diffusion because of the passivation layer, so stress relaxation is largely suppressed. The dislocation motion inside the Cu film is also constrained by the passivation. (In the case of Al, even the unpassivated film is naturally coated with a thin native oxide layer [29, 32].)

Figure 3.5 shows a representative experimental stress–temperature response of a passivated 400-nm thick Cu film, measured from the curvature method. The film was electron-beam-deposited on a quartz substrate and passivated with a 80-nm thick silicon oxide (SiO_2) layer [47]. Thin chromium interlayers were applied between Cu and the adjacent materials. After an initial heating to 450°C from the as-fabricated condition, a stabilized cyclic response between 20 and 450°C ensues, as shown in Fig. 3.5. After stabilization, the specimen was also immersed in liquid nitrogen and heated back, with the partial reheating curve included in the figure. Linear extrapolation of the cooling to and reheating from −196°C leads to a smooth

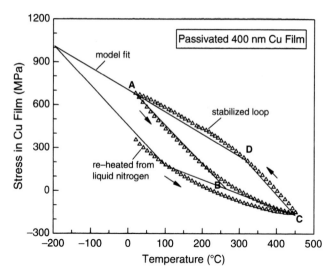

Fig. 3.5 Measured stress-temperature response of a passivated 400-nm thick Cu film [47]. A stabilized loop between 20 and 450°C is shown along with an interrupted partial cycle to −196°C. Experimental data are fitted with the kinematic hardening plastic model

connection of the measurable response, as evidenced by the "model fit" segments to be discussed below. Passivated Cu films with different thicknesses and/or deposition methods tested at various heating/cooling rates in different studies [29, 31, 35, 42, 47, 58], have revealed common features: rate-independent plasticity with the Bauschinger-like effect, i.e., early yielding upon load reversal (or even with negative yield stress such as at point B in Fig. 3.5). This behavior cannot be represented by the elastic-perfectly plastic model with temperature-dependent yield strength, because it will be impossible to endow the Cu film with a compressive yield strength that matches the heating response at elevated temperatures (e.g., from points B to C in the heating phase of the stabilized loop in Fig. 3.5). Post-yield *strain hardening* has to be involved.

Among the simple continuum plasticity models available, the kinematic hardening model, which assumes that the yield surface translates in stress space in the direction of outward normal [61–65] (see also Sect. 2.3), is a good candidate capable of describing the evolution of stress in passivated Cu films. The kinematic hardening model is able to predict the Bauschinger effect, and a steady state of alternating plastic deformation sets in after the initial loading cycle, which is compatible with the experimentally observed features.

Here, a useful procedure for extracting the temperature-dependent yield properties by invoking the kinematic hardening model with a linear hardening scheme is outlined [47]. A linear dependence of *initial* yield strength with temperature is assumed,

$$\sigma_y = \sigma_0 \cdot \left(1 - \frac{T}{T_0}\right), \tag{3.4}$$

where σ_y is the initial yield strength, T is temperature, and σ_0 and T_0 are reference constants. In Fig. 3.5, the stabilized loop between 20 and 450°C is highlighted with four points: A and C are the reversal points, and B and D roughly define the transitions from elastic to plastic deformations. Since segments AB and CD represent the elastic behavior (the stress state traverses diagonally inside the yield surface in the stress space), the two unknowns in (3.4), σ_0 and T_0, can then be determined by the following two equations

$$[\sigma_y]_A + [\sigma_y]_B = \left|[\sigma]_B - [\sigma]_A\right|,$$
$$[\sigma_y]_C + [\sigma_y]_D = \left|[\sigma]_D - [\sigma]_C\right|,$$

(3.5)

where $[\sigma_y]$ and $[\sigma]$ denote, respectively, the magnitude of yield strength at the specified temperature and the stress value at the specified point read from the plot (Fig. 3.5). The correspondence between the equations and the actual curve is depicted in Fig. 3.6, with the stress–temperature loop ABCD schematically representing that in Fig. 3.5. Note the left-hand sides of (3.5) represent the spans of the temperature-dependent elastic domain (yield surface), and the right-hand sides are the measured spans used for determining the yield strength values. The strain hardening rate (defined to be the slope of the uniaxial stress-strain curve beyond yielding) then follows

$$H_{D \to A} = \frac{\left|[\sigma]_A - [\sigma]_D\right| - \left\{[\sigma_y]_A - [\sigma_y]_D\right\}}{\Delta\alpha\Delta T},$$
$$H_{B \to C} = \frac{\left|[\sigma]_C - [\sigma]_B\right| - \left\{[\sigma_y]_B - [\sigma_y]_C\right\}}{\Delta\alpha\Delta T},$$

(3.6)

where $\Delta\alpha$ is the difference in CTE between Cu and the substrate, and ΔT is the temperature change involved. The solid curves shown in Fig. 3.5 are a result of the model with $\sigma_0 = 305$ MPa, $T_0 = 1{,}090$ K and $H = 77$ GPa. It satisfactorily fits the measured loop ABCD. With simple extrapolation to lower temperature, the model also matches the cryogenic thermal cycle quite well.

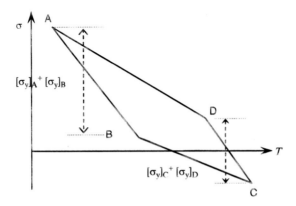

Fig. 3.6 Schematic illustrating the use of (3.5) for extracting the constants σ_0 and T_0 in (3.4). The stress–temperature loop ABCD represents a fit to the stabilized cycle measured in experiment

Table 3.1 Plastic yielding and hardening parameters of the Cu thin film extracted from the experimental stress–temperature curves [47]

Film thickness (nm)	σ_0 (MPa)	T_0 (K)	σ_y at 298 K (MPa)	H_{heat} (GPa)	H_{cool} (GPa)
400	305	1,090	222	77	77
250	337	1,652	276	89	101
125	350	3,303	318	110	124

The symbols H_{heat} and H_{cool} represent the hardening rates during heating and cooling, respectively

Table 3.2 Biaxial moduli obtained from the stress–temperature measurements of the Cu films [47]

Film thickness (nm)	Biaxial modulus during heating (GPa)	Biaxial modulus during cooling (GPa)
400	163.8	154.5
250	153.4	146.6
125	175.2	138.3
Average	164.1	146.5

The heating and cooling parts correspond to the beginning temperatures of 20 and 450°C, respectively, up to the respective apparent yielding. The coefficients of thermal expansion used for obtaining the biaxial moduli are: $\alpha_{quartz} = 0.5 \times 10^{-6}$ K^{-1} and $\alpha_{Cu} = 17.0 \times 10^{-6}$ K^{-1} (near room temperature) during heating; $\alpha_{quartz} = 0.74 \times 10^{-6}$ K^{-1} and $\alpha_{Cu} = 19.3 \times 10^{-6}$ K^{-1} (near 450°C) during cooling

In the analysis above, it is coincidental that the value H is the same during both heating (B to C) and cooling (D to A). The constitutive law does not require the two hardening rates to be equivalent. Table 3.1 lists the material constants of this 400-nm thick Cu film along with two other thinner Cu films obtained in the same set of experiments. It shows a significant size effect: the initial yield strength increases with decreasing film thickness. The strength of thinner film is also less temperature sensitive (greater T_0). In addition, the strain hardening rate also displays an increasing trend as the film thickness decreases. It is noticed that the values of H in all cases are quite high. The fact that passivated Cu films display strong strain hardening may be partially attributed to the confined small grains containing growth twins, because the twin boundaries can interact with dislocations and mediate the slip behavior [66–75]. The concept of "back stress," normally associated with the dislocation pile-up phenomenon as well as with the kinematic hardening model, is expected to play an important role in affecting the strain hardening response. Recent studies have also indicated the potential applicability of the back stress concept in dislocation slip behavior for constrained plastic flow in thin films [76–80]. It is also worth mentioning that the strong strain hardening in the present case is not at much variance with the tensile stress–strain response of Cu films measured mechanically [80–83]. When performing numerical simulation involving materials with significant strain hardening, it is imperative to incorporate the appropriate plastic model in the analysis. Further illustrations will be given in Chap. 4.

For completeness purpose, in Table 3.2 we list the biaxial modulus of the Cu film obtained from the same set of experiments using the same method described in Sect. 3.1.2. A moderate decrease in the elastic modulus with increasing temperature can be seen. It is also noted that there is no significant variation and clear trend for the different film thicknesses. This lack of size effect for the *elastic* behavior is typical of polycrystalline thin-film materials.

3.3 Film on a Compliant Substrate

There are circumstances that thin films are bonded to a much more compliant substrate. One application area is stretchable electronics, where small features of micro-components are deposited on a polymeric substrate [84–87]. On the more fundamental side, thin metal films are frequently deposited on a compliant substrate (such as polyimide) and the assembly is subject to mechanical testing to obtain the "free-standing" film behavior [5, 88–95]. An implicit assumption for this type of testing is that the substrate material is very compliant, so it tends not to "interfere" with the intrinsic deformation characteristics of the thin film. The stress–strain curve of the film can be determined by separating the force carried by the polymer substrate (through testing of the bare substrate) from that carried by the film/substrate structure. Alternatively, the strain and stress states in the metal film can be monitored in situ by X-ray diffraction [81].

In the above mentioned testing, the thickness of the polyimide substrate typically ranges from about 7 to 150 μm. While the substrate is much more compliant than the film, the true mechanical constraint it imposes on the film needs to be quantified. The approach considered in Fig. 3.2a can be applied but with the Si substrate replaced by polyimide, as shown in Fig. 3.7a. The 1-μm thick, elastic-perfectly plastic Al film is taken to have the same properties as before (Young's modulus 70 GPa, Poisson's ratio 0.33 and yield strength 200 MPa); the Young's modulus and Poisson's ratio of the elastic polyimide are 2.5 GPa and 0.34, respectively. Two polyimide thicknesses are considered: 7 and 125 μm. The simulated evolution of stresses σ_{xx}, σ_{yy} and the von Mises effective stress in the Al film as a function of applied strain ε_{xx}, under the condition of no edge effect during the tensile loading, is shown in Fig. 3.7b. Note the von Mises stress stays at 200 MPa once yielding commences. However, σ_{xx} continues to rise in the plastic regime as in the case of the Si substrate (Fig. 3.2 in Sect. 3.1.1). This again implies a "false" strain hardening behavior which will be measured in actual experiments. The change in σ_{xx} after yielding is larger for a thicker polyimide substrate. Additionally, a lateral tensile stress σ_{yy} also exists. When the Al film is still elastic, σ_{yy} is nearly zero due to the very close Poisson's ratios for Al (0.33) and polyimide (0.34). Upon yielding, this lateral stress component starts to develop and its magnitude can become quite significant if the substrate is sufficiently thick. (Note that this increasing lateral stress during plastic deformation has also been detected by X-ray diffraction in a 0.7-μm thick Cu film on a 125-μm thick polyimide substrate [81].) Therefore, care must be taken when interpreting experimentally obtained information, since a true macroscopic uniaxial stress state may *not* exist even with a compliant substrate.

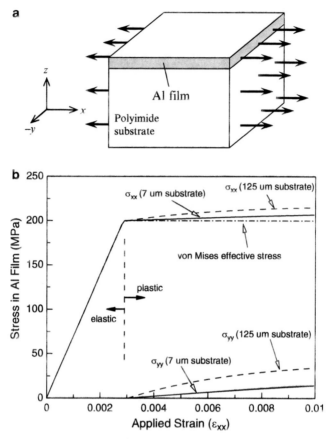

Fig. 3.7 (a) Schematic showing an Al film attached to a polyimide substrate. The entire structure is subject to uniform stretching (with prescribed boundary displacements). (b) The evolution of stresses σ_{xx}, σ_{yy}, and the von Mises effective stress inside the Al film for the two thicknesses of polyimide considered (The stress σ_{zz} is zero throughout the process)

In the case of 7-μm polyimide in Fig. 3.7b, the magnitude of σ_{yy} is significantly smaller. A thin substrate is thus desired for a better representation of the uniaxial stress state.

3.4 Mechanical Deflection of Microbeams

In certain MEMS-based test structures, a thin film is attached to a stiff substrate beam which is only moderately thicker than the film itself [3–5, 96]. A transverse deflection is imposed by mechanical means (such as by an instrumented indenter) and quantitative load-deflection information can be recorded. Figure 3.8a shows

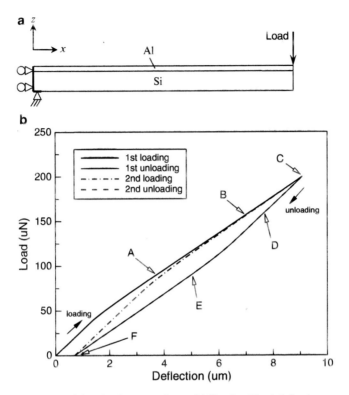

Fig. 3.8 (**a**) Schematic of the microbeam specimen. (**b**) Simulated load-deflection response of the microbeam. (**c**) Deformed configurations and contours of equivalent plastic strain corresponding to points A, B, C and F labeled along the curve in part (**b**) during the first cycle. (**d**) Deformed configurations and contours of stress component σ_{xx} corresponding to points A and C during the first loading and points D, E and F during the first unloading labeled along the curve in part (**b**)

a schematic of such a beam, consisting of an Al film on a Si substrate, to be deflected by an edge load. In the model the beam length, Al thickness and Si thickness are taken as 60, 1 and 5 μm, respectively. The loading and boundary conditions are also shown in Fig. 3.8a. For the purpose of demonstrating fundamental features first, a 2D plane stress model (no out-of-paper stress components) is used in the calculation. The influence of lateral constraint will be discussed at the end of this section.

Figure 3.8b shows the simulated load-deflection curve over a history of two full cycles between the applied loads of 0 and 200 N. The deflection is defined to be the downward vertical displacement of the point where the load is applied. It can be seen that, upon unloading from the peak load during the first cycle, the deflection does not return to zero when the load is completed removed. This is due to the plastic deformation in the film during loading so a residual deflection results. Starting from the second cycle, a stabilized load-deflection loop is established. Figure 3.8c shows the deformed configurations and contours of equivalent plastic

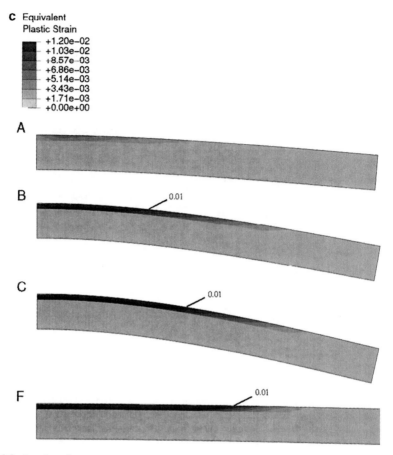

Fig. 3.8 (continued)

strain corresponding to points A, B, C and F labeled along the curve in Fig. 3.8b
during the first cycle. It is observed that, during the loading phase, plastic deformation in Al initiates near the support end of the beam and spreads towards the loading
end. The increasing bending moment from right to left is responsible for the non-
uniform stress distribution in the film. Unloading, however, is not merely an elastic
recovery process. Comparing the plastic strain fields at point C and F in Fig. 3.8c,
it is evident that further plastic deformation has occurred during the unloading
phase. Therefore this cyclic deflection test is fundamentally different from simpler
loading schemes such as typical cyclic uniaxial tests between zero and a peak
applied stress.

Figure 3.8d shows the contours of the stress component σ_{xx} corresponding to
points A, C, D, E and F labeled along the curve in Fig. 3.8b during the first cycle.
The variation of this predominant stress component throughout the beam structure
can be seen. Inside the Si substrate, tensile and compressive stresses develop in,

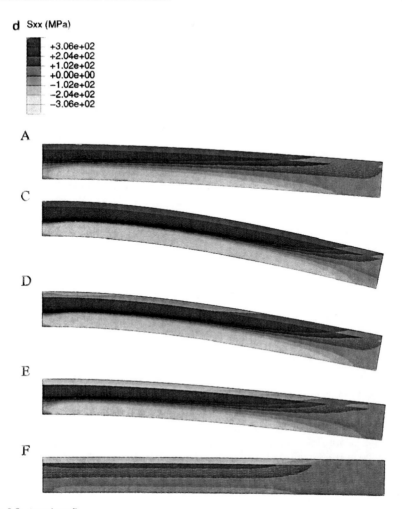

Fig. 3.8 (continued)

respectively, the upper and lower regions. For the elastic-plastic metal film, σ_{xx} reaches its peak magnitude (about 200 MPa) progressively from left to right during initial loading, as can be seen at points A and C. During unloading (D, E and F), stress relief occurs in a non-uniform manner. The stress decreases faster in regions closer to the left end of the film. At point E, reversed plasticity has already taken place and σ_{xx} in the left part of the film has reached the peak value (about -200 MPa) in compression. When the applied load drops to zero (point F), the majority of the film is under the peak compressive stress. Subsequent cycles will lead to a recurrent stress pattern and a hysteresis loop shown in Fig. 3.8b. It is thus realized that, in the microbeam deflection test considered, an applied cyclic load of the "zero-peak-zero" pattern actually results in a fully reversed "tension-compression" stress cycles inside the majority of the metal film.

Because the maximum bending moment occurs at the support (left) end of the structure during deformation, a uniform stress distribution in the metal film is not possible in typical microbeam specimens. To circumvent this limitation, a triangular beam configuration has been devised [96] which has the advantage that the entire beam is subject to a constant moment per unit width. This will result in a uniform state of strain on the top surface as the beam is deflected. It should be noted, however, that in addition to the varying stress along the axial direction, a through-thickness gradient also exists because of the relatively thin substrate. Moreover, in actual test structures the metal film is under a *plane strain* condition near the support end but away from it a *generalized plane strain* state prevails (allowing deformation in the out-of-paper direction). A finite lateral (out-of-paper direction) stress field is thus always present, which varies along the beam due to the changing lateral constraint. These problems cannot be avoided even with a "constant moment per width" scheme. As a consequence, care should be exercised when quantitatively analyzing the information obtained from the microbeam deflection experiment.

3.5   Thermal Deflection of Microbeams

3.5.1   Bi-layer Beam

In addition to mechanical bending considered in the previous section, deflection of microbeams can also be induced by the thermal expansion mismatch between the film and substrate materials when a temperature change is involved. Figure 3.9a shows the beam structure used in the present analysis. The thicknesses of the Al film and Si substrate are taken to be 1 and 4 μm, respectively. The total beam length is 60 μm. For simplicity a plane stress condition is again assumed in this case study. The left end of the beam is rigidly clamped, and the right end as well as the top and bottom surfaces are free. When a uniform temperature change (ΔT) is imposed, the mismatch in CTE between Si and Al will lead to bending of the beam. A stress-free condition is assumed before the first heating phase starts.

Figure 3.9b shows the evolution of deflection, defined to be the vertical displacement of the upper-right tip, during the first two full heating/cooling cycles over the ΔT range between 0 and 400°C, for the Si-Al microbeam. During initial heating the greater expansion of the Al film results in downward bending. At about $\Delta T = 230$°C, the deflection rate starts to decrease and a saturation value is soon reached. Upon cooling, a reduction in deflection takes place. When ΔT is decreased to about 160°C, the microbeam starts to bend upward so the deflection becomes positive. Therefore, the microbeam displays an opposite deflection and does not recover its original shape, although a temperature pattern of "zero-peak-zero" is imposed. This is due to plastic deformation in the Al film during heating (discussed below). The cyclic response stabilizes during the second cycle, which largely follows the thermoelastic behavior.

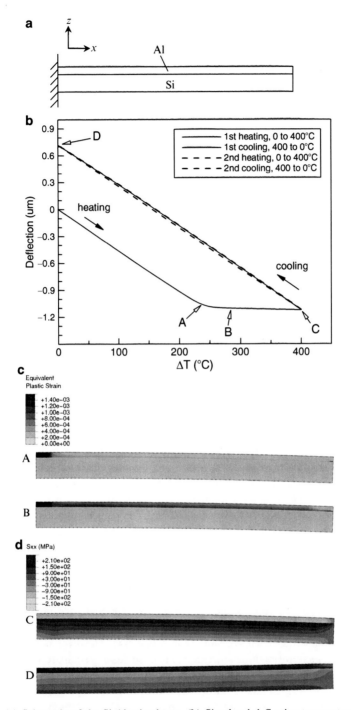

Fig. 3.9 (**a**) Schematic of the Si-Al microbeam. (**b**) Simulated deflection-temperature change (ΔT) response. (**c**) Deformed configurations and contours of equivalent plastic strain corresponding to points A and B labeled along the curve in part (**b**) during the first heating phase. (**d**) Deformed configurations and contours of stress component σ_{xx} corresponding to points C (end of the first heating phase) and D (end of the first cooling phase) labeled along the curve in part (**b**)

Figure 3.9c shows the deformed configurations and contours of equivalent plastic strain corresponding to points A and B labeled along the deflection-temperature curve in Fig. 3.9b during heating in the first cycle. At point A, except in the area near the rigidly clamped end where significant plastic deformation has occurred, yielding in the Al film has just started along the Al-Si interface. At point B essentially the entire Al film has yielded. A plastically deforming Al possesses the free-flowing capability to compensate for any further tendency for bending induced by thermal mismatch. Therefore the deflection stays constant afterwards. Figure 3.9d shows the contours of stress component σ_{xx} corresponding to points C (at the end of first heating) and D (at the end of the first cooling) labeled along the curve in Fig. 3.9b. At C, Al has yielded in compression and σ_{xx} is largely -200 MPa throughout. During cooling elastic unloading in Al first occurs, which is followed by the development of tensile axial stress. At D, σ_{xx} is still below 200 MPa in most regions of Al so reversed yielding has not initiated on a gross scale. If the applied temperature change chosen in this example is greater, reversed yielding can occur and a stabilized hysteresis loop will then appear during the second cycle.

3.5.2 Tri-layer Beam

We now consider a different case of the microbeam with a 1-μm thick silicon dioxide (SiO_2) passivation (capping) layer above the Al film, as schematically shown in Fig. 3.10a. Figure 3.10b shows the evolution of deflection during the first two full heating/cooling cycles over the ΔT range between 0 and 400°C. The deflection first progresses toward the negative side during heating. However, when ΔT is about 180°C, the bending reverses its direction and the magnitude of deflection starts to decrease. This means that, during the *monotonic* temperature change, a reversal in the deflection direction occurs. During the cooling phase of the first cycle, the deflection first advances toward the positive side. When ΔT is reduced to about 50°C, another reversal of the deflection direction takes place. The same features are then repeated in the subsequent cycle, leading to a distinct hysteresis loop.

The most interesting aspect revealed in Fig. 3.10b is the reversal of the deflection direction during a monotonic temperature variation. This is due to the plastic yielding in Al and the CTE relationship of the three materials involved (3.0×10^{-6} K^{-1} for Si, 23.0×10^{-6} K^{-1} for Al and 0.52×10^{-6} K^{-1} for SiO_2, assuming temperature independence). When the Al film is still elastic, its much higher tendency to expand plays a dominant role so a downward deflection results. However, when Al becomes fully plastic, its free-flowing nature prevents itself from participating in the thermal expansion "competition." In other words, the Al film essentially serves as a "glue" and only the CTE mismatch between Si and SiO_2 needs to be accounted for. The beam will thus tend to bend upward due to the high CTE of Si (compared to SiO_2), so a reversal takes place. Figure 3.10c shows the deformed configurations and contours of equivalent plastic strain corresponding to points A and B labeled along the deflection-temperature curve in Fig. 3.10b during the first heating phase. It is evident

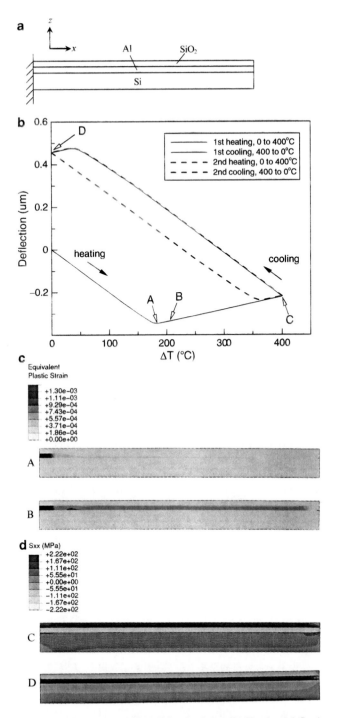

Fig. 3.10 (a) Schematic of the passivated Si-Al-SiO$_2$ microbeam. (b) Simulated deflection-temperature change (ΔT) response. (c) Deformed configurations and contours of equivalent plastic strain corresponding to points A and B labeled along the curve in part (b) during the first heating phase. (d) Deformed configurations and contours of stress component σ_{xx} corresponding to points C (end of the first heating phase) and D (end of the first cooling phase) labeled along the curve in part (b)

that the "turning point" is associated with the initiation of gross plasticity at the Al-Si interface in the Al film. When Al becomes fully plastic, the microbeam deflection is dictated by the thermoelastic response of Si and SiO_2. It is worth mentioning that the reversal will appear to be less abrupt if the Al film is thicker, because it will take a greater range of ΔT for plasticity to spread through the entire thickness of Al.

Figure 3.10d shows the contours of stress component σ_{xx} corresponding to points C (the end of the first heating) and D (the end of the first cooling) labeled along the curve in Fig. 3.10b. At C, a predominant compressive stress of about −200 MPa exists, while at D the tensile stress of about 200 MPa prevails because reversed plastic deformation has occurred. The returning phase of the temperature pattern of "zero-peak-zero" is again seen to result in stress evolution inside Al from the compressive side to well into the tensile side.

While numerical modeling is employed to illustrate fundamental features in the clamped microbeam during thermally induced deformation, it is noted that, for unsupported (free-standing) structures, analytical expressions for the development of internal stresses and overall beam curvature exist [97]. The analytical treatment also included the phenomenon of curvature reversal during monotonic temperature excursion, which is similar to the deflection reversal behavior presented above. It is worth mentioning that this concept may be further explored to design mechanisms for application in certain MEMS devices [98–100].

3.6   Rate-Dependent Behavior

The discussion of continuum-based models thus far has focused on a straightforward rate-independent elastic-plastic behavior. When a substrate-attached thin film is experiencing very slow thermal excursions or isothermal stress relaxation, a time-dependent constitutive model may need to be considered. An example is shown in Fig. 3.11 where the stress change with time in passivated Cu films, measured experimentally, during isothermal hold at different temperatures is seen [47]. In this case the material system is the same as that considered in Sect. 3.2 (Fig. 3.5): SiO_2-passivated 400-nm thick Cu film on a quartz substrate. The films were thermally cycled between 20 and 450°C to achieve stabilization and then isothermally held at various temperatures during the cooling phase. It can be seen from the thermal cycling result in Fig. 3.5 that, during cooling the stress in the Cu film is about zero at 400°C and becomes tensile at temperatures below 400°C. Figure 3.11 indeed shows that there is essentially no change in stress over time at 400°C. At lower temperatures, relaxation of the tensile stress occurs, with an initially higher stress resulting in a greater extent of relaxation during the isothermal hold. For instance, at 300°C the initial tensile stress is about 250 MPa, and it reduces to about 188 MPa at the end of 4 h.

To quantitatively describe the stress relaxation phenomenon, one may use the simple steady-state creep rate relation in (2.23),

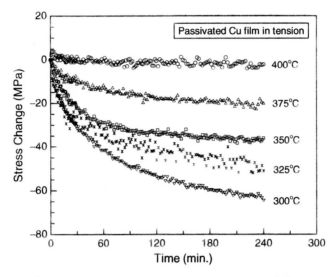

Fig. 3.11 Experimentally measured stress change as a function of time during isothermal hold of a passivated 400-nm thick Cu film. The beginning state was attained during cooling from 450°C, following the stress–temperature response shown in Fig. 3.5

$$\dot{\varepsilon}_s = A\sigma^{n_c} \exp\left(-\frac{Q}{RT}\right), \qquad (3.7)$$

and fit it to the experimental stress relaxation curves to determine the relevant material parameters. It was found that the parameters $A = 4.37 \times 10^{10}$ (MPa)$^{-10}$s^{-1}, $n_c = 10$ and $Q = 457$ kJ/mol may be used to fit the curves in Fig. 3.11 reasonably well. The stress exponent n_c and activation energy Q are considerably higher than those of bulk copper and are in the range characteristic of dislocation creep in dispersion or particle strengthened metals. This suggests that, within the stress and temperature ranges considered, *constrained* dislocation motion is the dominant creep mechanism in the passivated Cu film. The constraint effect apparently arises from the two interfaces bounding the thin material, which is qualitatively similar to the situation encountered in a metal matrix strengthened by dispersed particles [101].

3.6.1 The Deformation Mechanism Approach

While (3.7) together with the experimentally determined parameters define the rate-dependent material response, one should be aware that the same set of parameters may not be applied indiscriminately even for exactly the same material. This is because that, outside of the temperature and/or stress ranges used for obtaining the fitted parameters, the dominant deformation mechanism may not be the same so a different set of parameters may be required. Researchers in the thin-film materials community have attempted to account for rate-dependency by incorporating physics-based models relating the relaxation strain rate to stress and temperature using the generic form of (2.22). Below we adopt this "materials science" type of approach and summarize the governing equations of the individual deformation mechanisms, which will then be used for a case study on stress evolution in a Cu thin film during temperature cycles. The equations are taken from a compiled source on deformation mechanisms [102], but are presently organized into a form pertaining to the equi-biaxial stress state [103].

- *Dislocation glide.* This mechanism typically dominates at low temperatures under high stress. Although commonly treated as a rate-independent yielding phenomenon, dislocation glide is in fact a kinetic process and the strain rate can be given by

$$\dot{\varepsilon}_{dg} = \frac{\dot{\gamma}_0}{2\sqrt{3}} \exp\left[-\frac{\Delta F}{kT}\left(1 - \frac{\sigma}{\sqrt{3}\hat{\tau}} \right) \right], \tag{3.8}$$

  where $\dot{\gamma}_0$ is a material constant, ΔF is the activation energy for dislocation glide, $\hat{\tau}$ is the Mises flow strength of the material at 0 K, and k is Boltzmann's constant (1.381×10^{-23} J/K).

- *Power-law creep.* The physical mechanism responsible for power-law creep is climb-controlled dislocation glide, which typically becomes dominant at elevated temperatures. The strain rate can be expressed as

$$\dot{\varepsilon}_{pl} = \frac{A_{pl} D_{eff,pl} \mu b}{2kT} \left(\frac{\sigma}{\mu} \right)^n, \tag{3.9}$$

  where A_{pl} is the Dorn constant for power-law creep, μ is shear modulus, b is the magnitude of Burgers vector, n is the creep exponent with a value typically between 3 and 10, and

$$D_{eff,pl} = D_v \left[1 + \frac{10a_c}{b^2} \left(\frac{\sigma}{\sqrt{3}\mu} \right)^2 \frac{D_c}{D_v} \right], \tag{3.10}$$

  where D_v and D_c are the lattice diffusion coefficient and the dislocation core diffusion coefficient, respectively, and a_c is the cross-section area of the dislocation core along which fast diffusion takes place.

- *Power-law breakdown.* At higher stresses, the simple power law breaks down, where the measured strain rates are greater than those predicted by (3.9). This is a transition process from climb-controlled to glide-controlled flow with

$$\dot{\varepsilon}_{plb} = \frac{A_{pl}}{2\alpha'^n} \cdot \frac{D_{eff,pl} \mu b}{kT} \left[\sinh\left(\alpha' \frac{\sigma}{\mu} \right) \right]^n, \tag{3.11}$$

where α' is an empirical fitting parameter. Equation (3.11) reduces to (3.9) at about $\sigma < \mu / \alpha'$.

- *Diffusional flow.* In addition to deformation due to dislocation motion, a diffusive flux of matter can be induced by a stress field. The strain rate is expressed as

$$\dot{\varepsilon}_{\text{diff}} = \frac{7\sigma\Omega}{kTd^2} D_{\text{eff.diff}},$$
(3.12)

where Ω is the atomic volume, d is the average grain size, and

$$D_{\text{eff.diff}} = D_{\text{v}} \left[1 + \frac{\pi\delta}{d} \cdot \frac{D_{\text{b}}}{D_{\text{v}}} \right].$$
(3.13)

Here D_{b} is the grain boundary diffusion coefficient and δ is the effective thickness of the boundary. Lattice diffusion and grain boundary diffusion tend to control the creep rate at high and low temperatures, respectively.

All the above mechanisms are assumed to be able to contribute to the rate-dependent deformation. The net relaxation strain rate due to the current stress and temperature is then given by

$$\dot{\varepsilon}_{\text{rlx}} = \frac{d\varepsilon_{\text{rlx}}}{dt} = \text{greatest of } (\dot{\varepsilon}_{\text{dg}}, \dot{\varepsilon}_{\text{pl}}, \text{or } \dot{\varepsilon}_{\text{plb}}) + \dot{\varepsilon}_{\text{diff}}.$$
(3.14)

3.6.2 Stress Evolution During Temperature Cycles

We now present a case study on thin-film stress–temperature relationship during temperature cycles. The creep mechanisms outlined above are incorporated into the constitutive response of a thin copper film. The equi-biaxial stress in the film is caused by thermal expansion mismatch between the Cu film and the much thicker elastic quartz substrate. Note if the thin film behaves only elastically, the stress evolution will follow (3.2). Here the inelastic relaxation of strain $(\varepsilon_{\text{rlx}})$ needs to be accounted for

$$\sigma_f = \frac{E_f}{1 - v_f} \cdot \left[\left(\alpha_s - \alpha_f \right) \Delta T - \varepsilon_{\text{rlx}} \right].$$
(3.15)

The material parameters in (3.15) and the relaxation strain rates for all deformation mechanisms required as input are listed in Table 3.3. Note the stress calculation is analytical in nature but numerical iterations are needed to achieve convergence. The rate of temperature change (dT/dt) needs to be specified. At each time step $t=\Delta t$, $2\Delta t$, $3\Delta t$... (and thus the temperature change $(dT/dt)\Delta t$, $2(dT/dt)\Delta t$, $3(dT/dt)\Delta t$...), (3.14) and (3.15) are combined to solve for the change in relaxation strain, $\Delta\varepsilon_{\text{rlx}}$, iteratively until convergence. The stress σ_f can then be obtained. The numerical procedures are schematically outlined in Fig. 3.12.

Table 3.3 The material parameters used in the calculation of thin-film stress–temperature response taking into account the creep mechanisms [102]

	Cu	Quartz
α (°C^{-1})	$17 \times 10^{-6} + 7.0 \times 10^{-9}\,(T - 20)$	0.5×10^{-6}
μ (MPa)	$42100 - 16.765 \cdot (T - 20)$	–
ν	0.3	–
E (MPa)	$2(1 + \nu)\mu$	–
γ_0 (s^{-1})	10^{-6}	–
ΔF (J)	35.3×10^{-20}	–
$\hat{\tau}$ (MPa)	265	–
A_{pl}	7.4×10^{5}	–
b (m)	2.56×10^{-10}	–
n	4.8	–
D_v (m^2/s)	$2.0 \times 10^{-5} \exp[-197,000 / R(T + 273)]$	–
$a_c D_c$ (m^4/s)	$1.0 \times 10^{-24} \exp[-117,000 / R(T + 273)]$	–
α'	794	–
Ω (m^3)	1.18×10^{-29}	–
δD_b (m^3/s)	$5.0 \times 10^{-15} \exp[-104,000 / R(T + 273)]$	–
d (μm)	0.04	

All temperatures (T) are in °C. R is gas constant (8.134 J/mol)

Figure 3.13a shows the calculated stress in the Cu thin film as a function of temperature over a stabilized cycle between room temperature and 600°C. The result includes two different heating/cooling rates: 10 and 1°C/min. It can be seen that the film stress tends to relax nearly fully at elevated temperature. This is found to be in general agreement with experimental measurements of *unpassivated* Cu thin films attached to a substrate as mentioned in Sect. 3.2. A slower heating/cooling rate results in lower stress magnitudes, but the difference is quite small for the rates considered. The strong tendency for stress relaxation at high temperatures is dictated by the diffusional flow mechanism. To explore the effect of this particular creep mechanism, we intentionally "switch off" diffusional flow in the analysis while carrying out the simulation. The result is shown in Fig. 3.13b. Although still rate-dependent, the overall shape of stress–temperature curve now resembles the elastic-plastic type.

It should be noted, however, that the constitutive equations and material parameters used in the above example are based primarily on 3D polycrystalline materials with certain microstructural features. As a consequence they may not be representative of a given thin film material so the results in Fig. 3.13 should only be viewed as qualitative rather than quantitative. In an actual unpassivated thin film with a columnar grain structure, for instance, atomic diffusion along the free surface (rather than pure grain boundary and lattice diffusions) is perhaps the main relaxation mechanism at elevated temperatures. If the free surface is covered with a dense passivation layer, the diffusional process will become suppressed. Crudely speaking, the cases without and with passivation are qualitatively represented by

Fig. 3.12 Flow chart showing the numerical procedures employed in the thermal cycling analysis. Note that ε_{rlx} represents the accumulated relaxed strain up to the previous time step, and $\Delta\varepsilon_{rlx}$ the incremental relaxed strain during the current time step

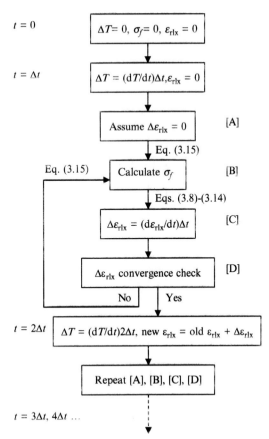

the stress–temperature behaviors in Fig. 3.13a and b, respectively. To this end it is worth mentioning that analytical models taking into account specific aspects of diffusional flow pertaining to the columnar grain structure have been reported [26, 41, 104–106]. More generally, the rate equations, in various formats with different levels of complexity, have been applied to investigate a wide variety of time-dependent mechanical response of polycrystalline blanket films bonded to stiff substrates [19, 24, 26, 27, 30, 31, 34, 38, 39, 41, 107–110]. One intriguing approach, similar to the methodology employed in Fig. 3.13 above, is that the quantitative relaxation models may be tuned for fitting the measured stress-temperature response [38], which can shed light on the relaxation mechanisms involved in actual thin-film systems.

As a final note, many time-dependent relaxation models, while more fundamental than the simple phenomenology, are still incompatible with some physical inelastic features. For instance, the commonly used power-law creep formulation (3.9) does

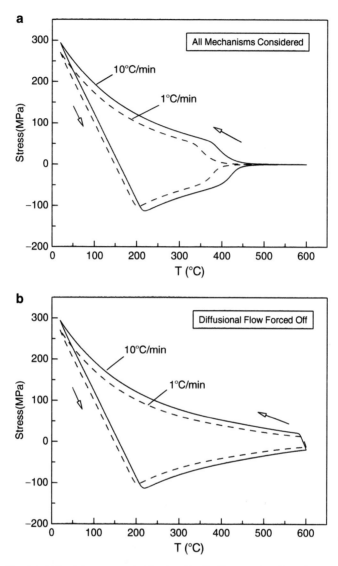

Fig. 3.13 Simulated bi-axial stress carried by the thin Cu film as a function of temperature over a stabilized cycle between 20 and 600°C, under two different heat/cooling rates: 10 and 1°C/min. In (**a**) all the possible deformation mechanisms are incorporated in the analysis; in (**b**) the "diffusional flow" mechanism is taken to be inactive

not involve a grain size parameter, and the many rate-independent cyclic stress–temperature responses simply cannot be recovered from the rate equations. Ideally there should be a *unique* set of constitutive models that, on one hand can give rise to accurate rate-dependent material response (such as in Fig. 3.11), and on the other can predict relatively rate-independent elastic-plastic response (such as in Fig. 3.5) under appropriate loading conditions over a sufficiently wide span of temperature, time and microstructural scales. Currently there is no such unified model available. As a consequence, when conducting deformation analyses one is advised to choose the "right" model that optimally suits the physical problem at hand.

3.7 Indentation Loading

The instrumented indentation technique has been widely used to probe mechanical properties, especially for materials with small dimensions [16, 111–115]. An indentation test requires minimal material preparation, and can be performed multiple times on a single specimen. It involves the measurement of applied load and penetration displacement (depth) when the indenter is pressed onto and retracted from the test material. The choice of applied load and indenter tip geometry enables the probing of different volumes and shapes of materials. The advancement of instrumentation has made possible the resolution limit down to piconewtons for force and nanometers for displacement. Nanoindentation has thus become essentially a standard method for obtaining basic mechanical properties of thin-film materials.

Nanoindentation is most commonly employed to measure the elastic modulus and hardness of a material. However, the deformation field underneath an indenter is complex even for a bulk homogeneous material so analysis of data is non-trivial. A typical load–displacement curve during indentation loading and unloading can be seen in Fig. 3.14b. The elastic modulus of the material is normally obtained according to the Oliver and Pharr method [111], which is based on the expression

$$S = \beta \frac{2}{\sqrt{\pi}} E_r \sqrt{A}, \tag{3.16}$$

where S is the contact stiffness defined to be the initial unloading slope of the load–displacement curve, A is the contact area projected onto the plane of the original specimen surface at the onset of unloading, β is a geometry-dependent dimensionless parameter close to unity, and E_r is the reduced modulus given by

$$\frac{1}{E_r} = \frac{1-v^2}{E} + \frac{1-v_i^2}{E_i}. \tag{3.17}$$

In (3.17) E and v are the Young's modulus and Poisson's ratio, respectively, of the material being tested, and E_i and v_i are the Young's modulus and Poisson's ratio, respectively, of the indenter. If β, E_i, v_i and v are known (or assumed), the Young's modulus can then be obtained from (3.16) and (3.17) based on the measured S and A at the prescribed peak indentation load (or depth). The hardness of the material, H, is directly obtained from

$$H = \frac{P}{A}, \tag{3.18}$$

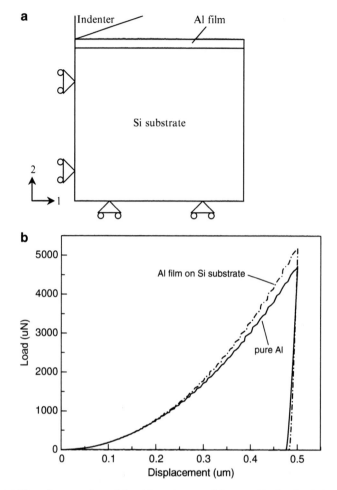

Fig. 3.14 (a) The axisymmetric model and boundary conditions used in the simulation of loading by a conical indenter. The left boundary is the symmetry axis. (b) Representative indentation load–displacement curves obtained from the modeling. The case of Al film on the Si substrate results in a higher indentation load at a given displacement, compared to the case of pure Al (a homogeneous Al block)

where P is the applied load. In experiments the projected contact area A is not easily measured. Procedures have been developed to indirectly determine A from the directly measured load–displacement response [111–113]. When performing numerical indentation analyses using the finite element method, the last nodal point on the top surface in contact with the indenter can be identified in the deformed mesh so the effect of possible "pile-up" or "sink-in" of material at the indentation edge can be automatically taken into account.

Applying nanoindentation to thin films has now become a routine operation. It is recognized, however, that the underlying substrate material can make a significant contribution to the deformation so the measured property may not be true to the thin film of concern. A common practice is then to apply indentation to a depth within 10% of the film thickness, with the perception that the indentation induced deformation is sufficiently far from the interface so any substrate effect can be ignored [111]. In the following we utilize a numerical finite element analysis to test this assumption. It also serves to illustrate how the thin film deformation is constrained on both sides by the indenter and substrate materials.

Figure 3.14a shows a schematic of the model and the boundary conditions used in the modeling. The model is axisymmetric and features a conical indenter with a semi-angle of 70.3°, which results in the same contact area as the Berkovich indenter in typical nanoindentation experiments. Use of the conical indenter is a practical way to model this type of indentation loading in a 2D configuration [111]. A 1-μm thick Al thin film is attached to a 100-μm thick Si substrate. The width (radius) of the model is 100 μm. The left boundary (symmetry axis) is allowed to move only in the 2-direction. The bottom boundary is allowed to move only in the 1-direction. The right boundary is not constrained. The top boundary, when not in contact with the indenter, is also free to move. The coefficient of friction between Al and the diamond indenter is taken to be 0.1. The input properties of Al and Si are taken to be the same as those in Sects. 3.1, 3.4 and 3.5 ($E_{Al} = 70$ GPa, $v_{Al} = 0.33$, $\sigma_{y, Al} = 200$ MPa, $E_{Si} = 130$ GPa and $v_{Si} = 0.28$). The Young's modulus and Poisson's ratio of the elastic diamond indenter are 1,141 GPa and 0.07, respectively. For comparison purposes we also consider indentation on a pure Al block without substrate. In this case the entire Si substrate in the model is replaced with Al. In the simulation the indenter displacement is incrementally prescribed. The resulting load is then the reaction force along the 2-direction acting on the indenter.

Figure 3.14b shows two representative load–displacement curves obtained from the modeling, one of the pure Al block and one of the Al thin film on the Si substrate. The maximum indentation depth, 0.5 μm, is at one half of the Al film thickness. The case with the Si substrate shows higher loads than the pure Al model, indicating the influence of the substrate material on the indentation response.

3.7.1 Indentation-Derived Elastic Modulus and Hardness

Attention is now turned to the elastic modulus and hardness of the Al thin film (on a Si substrate). Calculations of the modulus and hardness values follow (3.16)–(3.18). When obtaining the Young's modulus E from the reduced modulus in (3.17), the Poisson's ratio of 0.33 is used. Figure 3.15 shows the indentation-derived Young's modulus and hardness of the Al film as a function of indentation depth. The maximum depth considered is 0.6 μm, which corresponds to 60% of the initial film thickness. The Young's modulus used in the model input is 70 GPa. It can be seen

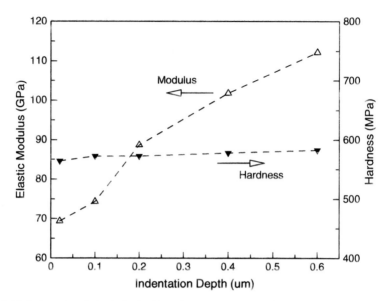

Fig. 3.15 Indentation-derived elastic (Young's) modulus and hardness, as a function of indentation depth, for the Al thin film attached to a thick Si substrate. The Young's modulus and yield strength of the elastic-perfectly plastic Al used as model input are, respectively, 70 GPa and 200 MPa

that this modulus value can be recovered from the indentation test only when the indentation depth is very small, namely below about 2% of the film thickness. The measured modulus (from indentation modeling) increases significantly as the indentation depth increases. At depths of 10 and 20% of the film thickness, the errors reach about 7 and 27%, respectively. Therefore a very strong effect of the substrate is present (note the Young's modulus of Si used as model input is 130 GPa). The hardness, on the other hand, behaves very differently. It is seen in Fig. 3.15 that the hardness value stays nearly constant, even when the indentation depth exceeds 50% of the film thickness. Eventually the hardness will increase significantly as the indenter tip moves closer to the interface, but it is apparent the substrate plays a much smaller role in affecting the hardness value compared to the case of elastic modulus. As a consequence, the "safe" range of indentation depth for probing the hardness of thin films is much greater. It is worth mentioning that the yield strength used as model input for the elastic-perfectly plastic Al is 200 MPa, so the ratio between the hardness and yield strength is around 2.86 according to Fig. 3.15.

The very different influences of the substrate on the measured properties can be explained by considering the nature of the indentation-derived quantities. In the case of elastic modulus, it is the recovery of deformation at the beginning of unloading that determines the contact stiffness. The thin film and the vast volume of the substrate both participate in the elastic rebound process. A stiffer substrate undergoes lesser deformation during loading and therefore makes a smaller contri-

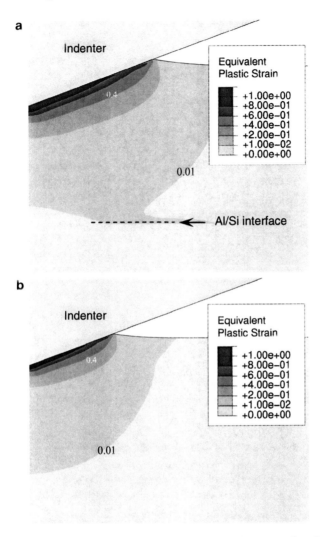

Fig. 3.16 Contour plots of equivalent plastic strain when the indentation depth reaches 0.25 μm for (**a**) the model with a 1 μm thick Al film on a Si substrate and (**b**) the model with a pure Al block

bution to the recovered displacement upon unloading, thus leading to a higher contact stiffness. The hardness, however, is dictated by the plastic yielding response. The size of the plastic zone in the thin film is therefore an important factor. Figure 3.16a shows a contour plot of equivalent plastic strain when the indentation depth is at 0.25 μm. For comparison the corresponding plot for the pure Al model is shown in Fig. 3.16b. It can be seen in Fig. 3.16a that, at the present indentation depth corresponding to 25% of the film thickness, the distribution of plastic strain

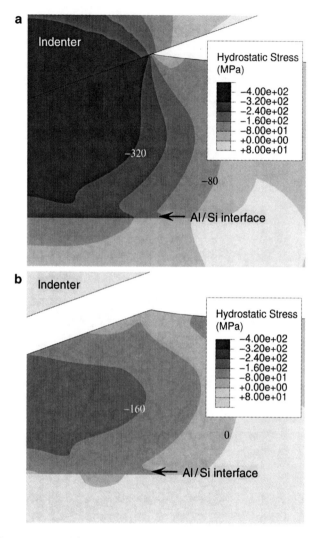

Fig. 3.17 Contour plots of hydrostatic stress in the Al film/Si substrate model in the cases (**a**) when the indentation depth reaches 0.25 μm and (**b**) after unloading

is affected by the substrate only slightly (in the figure only the lightest shade, corresponding to the plastic strain value of 0.01, is affected). Practically speaking, the plastic zone size in the Al-on-Si model is nearly the same as that in the all-Al model. As a consequence, the influence from the Si substrate is still very small.

Figure 3.17a shows the contour plot of hydrostatic stress generated in the Al film/Si substrate model when the indentation depth is at 0.25 μm. Although the yield strength of Al is 200 MPa, significantly greater magnitudes of triaxial compression is seen to exist underneath the indenter. High stress gradients exist so the stresses are much reduced in regions away from the indentation. It is important to note that, after unloading, a residual stress field remains in the material as observed in Fig. 3.17b. This is caused by the plastic deformation in the Al film during the loading phase. The highest magnitude of the residual compressive stress is still significantly greater than 160 MPa. It is also apparent from the presentation in Figs. 3.16 and 3.17 that pile-up of the Al material at the edge of the indentation has occurred, which is typical of materials possessing a low or zero strain hardening response.

The example above illustrated the constraint imposed on the thin film due to the indenter and substrate. It should be noted that the quantitative results are valid only for the specific material system considered. Further numerical analyses need to be conducted to characterize the substrate effect for different combinations of the film and substrate materials.

3.8 Projects

1. Consider the numerical example in Fig. 3.1. Carry out finite element modeling of the same kind of problem with the elastic-plastic Al thin film. Choose a point in the inner region of Al (close to the symmetry axis), and plot its stress value σ_{rr} as a function of the applied temperature change for two complete cycles (one cycle means cooling and then heating back to the initial temperature). If a layer of an elastic material (for instance, SiO_2) of the same thickness as Al is added on top of the Al film in the model, will the result be affected? How will the detailed stress pattern (contours) in Al and Si be affected? Discuss your observation.

2. Construct a numerical model of a thin film on a thick substrate (for example, a 1-μm thick Al film on a 200-μm thick Si substrate). The axisymmetric approach delineated in Fig. 3.1 may be used, but the in-plane dimension (r in Fig. 3.1) must be much greater than the substrate thickness. Impose a spatially uniform temperature change on the model, and observe the development of stress and curvature, focusing on regions away from the free edge of the film. Verify that (3.2) is followed. Verify that the Stoney's equation, (3.3), is followed. Are you able to generate the stress-temperature loop shown in Fig. 3.4 if the temperature-dependent yield strength of Al is used in the model?

3. Construct a numerical model of a thin film on a thick substrate, similar to the one in Project 2 above. Apply a spatially uniform temperature change of several cycles of cooling and heating, and obtain the evolution of stress in the film as a function of temperature (similar to the stress-temperature loops in Figs. 3.4 and 3.5). Start with the assumption that the film material is elastic-perfectly plastic. Then, explore the variation of the stress–temperature response affected by the additional constitutive features of the film material: Incorporating different extents of linear strain hardening and compare the results using the isotropic hardening model and the kinematic hardening model. What if the temperature-dependent initial yield strength and/or strain hardening rate are used?

4. Consider the tensile stretching problem in Fig. 3.7. Construct a similar model and simulate the cyclic response of the thin film (for example, with a strain range between −0.01 and 0.01). Observe the evolution of σ_{xx}, σ_{yy}, von Mises effective stress and equivalent plastic strain for several cycles, with and without the strain hardening of the thin film accounted for in the model. Will the isotropic hardening model and kinematic hardening model generate different responses? Discuss the practical implications, if any, of your findings.

5. Construct a finite element model of a film/substrate microbeam, as in Fig. 3.8. Carry out the mechanical deflection analysis using the plane stress model as in Fig. 3.8. In addition, carry out the same calculation but using the plane strain as well as the generalized plane strain models. Compare their overall load-deflection response and the development of stress and strain fields. What will the results look like if the strain hardening behavior is included in the thin film?

6. With reference to the discussion in Sect. 3.5, identify material and thickness combinations, other than the Si-Al-SiO$_2$ example, for microbeams that can display reversal in deflection direction during a monotonic temperature variation. Will there be any general trend that can be established for predicting such a behavior? Is it possible to design beams that can show even more complex deflection response? Explore this idea.

7. Consider the thin film/substrate model depicted in Fig. 3.1. Assume now that the Al film follows a rate-dependent response in the form of (3.7). Choose a constitutive model (may be a creep model) for Al found in the literature, and carry out a finite element analysis of thermal cooling from a zero-stress state. Examine the evolution of stress and strain fields during the process. How will the results differ if you apply different cooling rates? Is there a limit so that a rate-independent elastic-plastic model can yield a same response?

8. With reference to the discussion in Sect. 3.5, apply a rate-dependent model for the metal layer and numerically study the evolution of beam deflection affected by heating and cooling rates. Will there be any reversal of the deflection direction during a monotonic temperature change? What will occur if you insert certain isothermal hold period during the thermal cycling process? Correlate the internal stress and strains fields with the overall beam curvature and discuss your findings.

9. Assume that a thin-film material following the creep response of (3.7) is attached to a substrate. Due to the CTE mismatch stress is built up in the film during temperature cycles. Apply the same approach of numerical iteration as in Sect. 3.6.2 to calculate the stress–temperature response. (Here (3.7) is the only "deformation mechanism.") Adjust the parameters in (3.7) and examine the shape change of the curve. Are you able to correlate the simulation result with some experimental measurements on actual metallic films, and under what circumstances?

10. With reference to the numerical procedures considered in Sect. 3.6.2, conduct a literature survey and substitute some constitutive models with those taking into account the thin film microstructure (for example, diffusional flow for a columnar grain structure). Carry out a few cases of simulations and observe the overall stress–temperature response. Make comparisons with experimental results available in the literature.

11. With reference to the nanoindentation simulation in Sect. 3.7, conduct a series of finite element analysis assuming a free-sliding contact and various coefficients of friction between the indenter and the material being indented. How will the overall load–displacement curve be affected by the contact friction? How will the indentation-derived elastic modulus and hardness be affected by the contact friction? How will the stress and strain fields be affected?

12. With reference to the nanoindentation simulation in Sect. 3.7, conduct a series of finite element analysis assuming different input elastic properties and plastic properties of the thin-film material. For the variation of plastic properties, include different initial yield strengths as well as strain hardening behaviors. How will the indentation-derived elastic modulus and hardness be affected? How strong will the substrate effect be at different indentation depths? Also observe the possible pile-up or sink-in profiles at the edge of the indentation.

13. With reference to the nanoindentation simulation in Sect. 3.7, conduct a series of finite element analysis using the same Al thin film but with different substrate properties. Vary the Young's modulus of the substrate from below that of Al to an extremely high value (approaching the case of a rigid substrate). How will the indentation-derived elastic modulus and hardness of the thin film be affected? How strong will the substrate effect be at different indentation depths?

References

1. I. S. Sokolnikoff (1956) Mathematical theory of elasticity, 2nd ed., McGraw-Hill, New York.
2. S. P. Timosheko and J. N. Goodier (1970) Theory of elasticity, 3rd ed., McGraw-Hill, New York.
3. T. P. Weihs, S. Hong, J. C. Bravman and W. D. Nix (1988) "Mechanical deflection of cantilever microbeams: A new technique for testing the mechanical properties of thin films," Journal of Materials Research, vol. 3, pp. 931–942.
4. S. P. Baker and W. D. Nix (1994) "Mechanical properties of compositionally modulated Au-Ni thin films – nanoindentation and microcantilever deflection experiments," Journal of Materials Research, vol. 9, pp. 3131–3145.
5. R. Schwaiger, G. Dehm and O. Kraft (2003) "Cyclic deformation of polycrystalline Cu films," Philosophical Magazine, vol. 83, pp. 693–710.
6. G. G. Stoney (1909) "The tension of metallic films deposited by electrolysis," Proceedings of Royal Society (London), vol. A82, pp. 172–175.
7. T. F. Retajczyk and A. K. Sinha (1980) "Elastic stiffness and thermal expansion coefficient of BN films," Applied Physics Letters, vol. 36, pp. 161–163.
8. J. T. Pan and I. A. Blech (1984) "In situ measurement of refractory silicides during sintering," Journal of Applied Physics, vol. 55, pp. 2874–2880.
9. P. A. Flinn, D. S. Gardner and W. D. Nix (1987) "Measurement and interpretation of stress in aluminum-based metallization as a function of thermal history," IEEE Transactions on Electron Devices, vol. ED34, pp. 689–699.
10. C. A. Volker (1991) "Stress and plastic flow in silicon during amorphization by ion-bombardment," Journal of Applied Physics, vol. 70, pp. 3521–3527.
11. A. L. Shull and F. Spaepen (1996) "Measurements of stress during vapor deposition of copper and silver thin films and multilayers," Journal of Applied Physics, vol. 80, pp. 6243–6256.
12. J. A. Floro and E. Chason (1996) "Measuring Ge segregation by real-time stress monitoring during $Si_{1-x}Ge_x$ molecular beam epitaxy," Applied Physics Letters, vol. 69, pp. 3830–3832.

13. A. J. Rosakis, R. B. Singh, Y. Tsuji, E. Kolawa and N. R. Moore (1998) "Full field measurements of curvature using coherent gradient sensing – application to thin-film characterization," Thin Solid Films, vol. 325, pp. 42–54.

14. A. E. Giannakopoulos, I. A. Blech and S. Suresh (2001) "Large deformation of layered thin films and flat panels: effects of gravity," Acta Materialia, vol. 49, pp. 3671–3688.

15. M. F. Doerner, D. S. Gardner and W. D. Nix (1986) "Plastic properties of thin films on substrates as measured by submicron indentation hardness and substrate curvature techniques," Journal of Materials Research, vol. 1, pp. 845–851.

16. W. D. Nix (1989) "Mechanical properties of thin films," Metallurgical Transactions A, vol. 20A, pp. 2217–2245.

17. A. K. Sinha and T. T. Sheng (1978) "The temperature dependence of stresses in aluminum films on oxidized silicon substrates," Thin Solid Films, vol. 48, pp. 117–126.

18. M. Hershkovitz, I. A. Blech and Y. Komem (1985) "Stress relaxation in thin aluminum films," Thin Solid Films, vol. 130, pp. 87–93.

19. V. M. Koleshko, V. F. Belitsky and I. V. Kiryushin (1986) "Stress relaxation in thin aluminum films," Thin Solid Films, vol. 142, pp. 199–212.

20. S. T. Chen, C. H. Yang, F. Faupel and P. S. Ho (1988) "Stress relaxation during thermal cycling in metal/polyimide layered films," Journal of Applied Physics, vol. 64, pp. 6690–6698.

21. D. S. Gardner and P. A. Flinn (1988) "Mechanical stress as a function of temperature in aluminum films," IEEE Transactions on Electron Devices, vol. 35, pp. 2160–2169.

22. D. S. Gardner and P. A. Flinn (1990) "Mechanical stress as a function of temperature for aluminum alloy films," Journal of Applied Physics, vol. 67, pp. 1831–1844.

23. R. Venkatraman, J. C. Bravman, W. D. Nix, P. W. Davis, P. A. Flinn and D. B. Fraser (1990) "Mechanical properties and microstructural characterization of Al-0.5%Cu thin films," Journal of Electronic Materials, vol. 19, pp. 1231–1237.

24. P. A. Flinn (1991) "Measurement and interpretation of stress in copper films as a function of thermal history," Journal of Materials Research, vol. 6, pp. 1498–1501.

25. R. Venkatraman and J. C. Bravman (1992) "Separation of film thickness and grain boundary strengthening effects in Al thin films on Si," Journal of Materials Research, vol. 7, pp. 2040–2048.

26. M. D. Thouless, J. Gupta and J. M. E. Harper (1993) "Stress development and relaxation in copper films during thermal cycling," Journal of Materials Research, vol. 8, pp. 1845–1852.

27. C. A. Volkert, C. F. Alofs and J. R. Liefting (1994) "Deformation mechanisms of Al films on oxidized Si wafers," Journal of Materials Research, vol. 9, pp. 1147–1155.

28. S. Bader, E. M. Kalaugher and E. Arzt (1995) "Comparison of mechanical properties and microstructgure of Al(1 wt.%Si) and Al(1 wt.%Si, 0.5 wt.%Cu) thin films," Thin Solid Films, vol. 263, pp. 175–184.

29. R. P. Vinci, E. M. Zielinski and J. C. Bravman (1995) "Thermal strain and stress in copper thin-films," Thin Solid Films, vol. 262, pp. 142–153.

30. Y.-L. Shen and S. Suresh (1995) "Thermal cycling and stress relaxation response of Si-Al and Si-Al-SiO2 layered thin films," Acta Metallurgica et. Materialia, vol. 43, pp. 3915–3926.

31. M. D. Thouless, K. P. Rodbell and C. Cabral, Jr. (1996) "Effect of a surface layer on the stress relaxation of thin films," Journal of Vacuum Science and Technology A, vol. 14, pp. 2454–2461.

32. R. P. Vinci and J. J. Vlassak (1996) "Mechanical behavior of thin films," Annual Review of Materials Science, vol. 26, pp. 431–462.

33. I.-S. Yeo, S. G. H. Anderson, D. Jawarani, P. S. Ho, A. P. Clark, S. Saimoto, S. Ramaswami and R. Cheung (1996) "Effects of oxide overlayer on thermal stress and yield behavior of Al alloy films," Journal of Vacuum Science and Technology B, vol. 14, pp. 2636–2644.

34. J. Proost, A. Witvrouw, P. Cosemans, Ph. Roussel and K. Maex (1997) "Stress relaxation in Al(Cu) thin films," Microelectronic Engineering, vol. 33, pp. 137–147.

35. R.-M. Keller, S. P. Baker and E. Arzt (1998) "Quantitative analysis of strengthening mechanisms in thin Cu films: Effects of film thickness, grain size, and passivation," Journal of Materials Research, vol. 13, pp. 1307–1317.

36. Y.-L. Shen, S. Suresh, M. Y. He, A. Bagchi, O. Kienzle, M. Ruhle and A. G. Evans (1998) "Stress evolution in passivated thin films of Cu on silica substrates," Journal of Materials Research, vol. 13, pp. 1928–1937.

37. J. Koike, S. Utsunomiya, Y. Shimoyama, K. Maruyama and H. Oikawa (1998) "Thermal cycling fatigue and deformation mechanism in aluminum alloy thin films on silicon," Journal of Materials Research, vol. 13, pp. 3256–3264.

38. R.-M. Keller, S. P. Baker and E. Arzt (1999) "Stress-temperature behavior of unpassivated thin copper films," Acta Materialia, vol. 47, pp. 415–426.

39. A. Witvrouw, J. Proost, Ph. Roussel, P. Cosemans and K. Maex (1999) "Stress relaxation in Al-Cu and Al-Si-Cu thin films," Journal of Materials Research, vol. 14, pp. 1246–1254.

40. M. J. Kobrinsky and C. V. Thompson (2000) "Activation volume for inelastic deformation in polycrystalline Ag thin films," Acta Materialia, vol. 48, pp. 625–633.

41. D. Weiss, H. Gao and E. Arzt (2001) "Constrained diffusional creep in UHV-produced copper thin films," Acta Materialia, vol. 49, pp. 2395–2403.

42. R. P. Vinci, S. A. Forrest and J. C. Bravman (2002) "Effect of interface conditions on yield behavior of passivated copper thin films," Journal of Materials Research, vol. 17, pp. 1863–1870.

43. G. Dehm, T. Wagner, T. J. Balk and E. Arzt (2002) "Plasticity and interfacial dislocation mechanisms in epitaxial and polycrystalline Al films constrained by substrates," Journal of Materials Science and Technology, vol. 18, pp. 113–117.

44. G. Dehm, T. J. Balk, H. Edongue and E. Arzt (2003) "Small-scale plasticity in thin Cu and Al films," Microelectronic Engineering, vol. 70, pp. 412–424.

45. S. P. Baker, R.-M. Keller-Flaig and J. B. Shu (2003) "Bauschinger effect and anomalous thermomechanical deformation induced by oxygen in passivated thin Cu films on substrates," Acta Materialia, vol. 51, pp. 3019–3036.

46. T. J. Balk, G. Dehm and E. Arzt (2003) "Parallel glide: unexpected dislocation motion parallel to the substrate in ultrathin copper films," Acta Materialia, vol. 51, pp. 4471–4485.

47. Y.-L. Shen and U. Ramamurty (2003) "Constitutive response of passivated copper films to thermal cycling," Journal of Applied Physics, vol. 93, pp. 1806–1812.

48. T. K. Schmidt, T. J. Balk, G. Dehm and E. Arzt (2004) "Influence of tantalum and silver interlayers on thermal stress evolution in copper thin films on silicon substrates," Scripta Materialia, vol. 50, pp. 733–737.

49. S. Hyun, O. Kraft and R. P. Vinci (2004) "Mechanical behavior of Pt and Pt-Ru solid solution alloy thin films," Acta Materialia, vol. 52, pp. 4199–4211.

50. P. Wellner, G. Dehm, O. Kraft and E. Arzt (2004) "Size effect in the plastic deformation of NiAl thin films," Zeitschrift für Metallkunde, vol. 95, pp. 769–778.

51. Y. Sun, J. Ye, Z. Shan, A. M. Minor and T. J. Balk (2007) "The mechanical behavior of nanoporous gold thin films," JOM, vol. 59(9), pp. 54–58.

52. P. A. Flinn and G. A. Waychunas (1988) "A new x-ray diffraction design for thin-film texture, strain, and phase characterization," Journal of Vacuum Science and Technology B, vol. 6, pp. 1749–1755.

53. M. F. Doerner and S. Brennan (1988) "Strain distribution in thin aluminum films using x-ray depth profiling," Journal of Applied Physics, vol. 63, pp. 126–131.

54. M. A. Korhonen, C. A. Paszkiet, R. D. Black and C.-Y. Li (1990) "Stress relaxation of continuous film and narrow line metallizations of aluminum on silicon substrates," Scripta Metallurgica et. Materialia, vol. 24, pp. 2297–2302.

55. P. A. Flinn and C. Chiang (1990) "X-ray diffraction determination of the effect of various passivations on stress in metal films and patterned lines," Journal of Applied Physics, vol. 67, pp. 2927–2931.

56. I. C. Noyan, J. Jordan-Sweet, E. G. Liniger and S. K. Kaldor (1998) "Characterization of substrate/thin-film interfaces with x-ray microdiffraction," Applied Physics Letters, vol. 72, pp. 3338–3340.

57. O. Kraft, M. Hommel and E. Arzt (2000) "X-ray diffraction as a tool to study the mechanical behavior of thin films," Materials Science and Engineering A, vol. 288, pp. 209–216.

58. S. P. Baker, A. Kretschmann and E. Arzt (2001) "Thermomechanical behavior of different texture components in Cu thin films," Acta Materialia, vol. 49, pp. 2145–2160.
59. T. Hanabusa, K. Kusaka and O. Sakata (2004) "Residual stress and thermal stress observation in thin copper films," Thin Solid Films, vol. 459, pp. 245–248.
60. E. Eiper, J. Keckes, K. J. Martinschitz, I. Zizak, M. Cabie and G. Dehm (2007) "Size independent stresses in Al thin films thermally strained down to −100°C," Acta Materialia, vol. 55, pp. 1941–1946.
61. A. Mendelson (1968) Plasticity: theory and application, MacMillan, New York.
62. R. Hill (1950) The mathematical theory of plasticity, Oxford University Press, Oxford.
63. J. Lubliner (1990) Plasticity theory, Macmillan, New York.
64. S. Suresh (1998) Fatigue of Materials, 2nd ed., Cambridge University Press, Cambridge.
65. T. J. Chung (2007) General continuum mechanics, Cambridge University Press, Cambridge.
66. J. P. Hirth and J. Lothe (1982) Theory of dislocations, 2nd ed., Wiley-Interscience, New York.
67. J. W. Christian and S. Mahajan (1995) "Deformation twinning," Progress in Materials Science, vol. 39, pp. 1–157.
68. A. J. Cao, Y. G. Wei and S. X. Mao (2007) "Deformation mechanisms of face-centered-cubic metal nanowires with twin boundaries," Applied Physics Letters, vol. 90, 151909.
69. T. Zhu, J. Li, A. Samanta, H. G. Kim and S. Suresh (2007) "Interfacial plasticity governs strain rate sensitivity and ductility in nanostructured metals," Proceedings of the National Academy of Sciences of the United States of America, vol. 104, pp. 3031–3036.
70. M. Dao, L. Lu, Y. F. Shen and S. Suresh (2006) "Strength, strain-rate sensitivity and ductility of copper with nanoscale twins," Acta Materialia, vol. 54, pp. 5421–5432.
71. L. Lu, R. Schwaiger, Z. W. Shan, M. Dao, K. Lu and S. Suresh (2005) "Nano-sized twins induce high rate sensitivity of flow stress in pure copper," Acta Materialia, vol. 53, pp. 2169–2179.
72. Z. H. Jin, P. Gumbsch, E. Ma, K. Albe, K. Lu, H. Hahn and H. Gleiter (2006) "The interaction mechanism of screw dislocations with coherent twin boundaries in different face-centered cubic metals," Scripta Materialia, vol. 54, pp. 1163–1168.
73. M. Dao, L. Lu, R. J. Asaro, J. T. M. De Hosson and E. Ma (2007) "Toward a quantitative understanding of mechanical behavior of nanocrystalline metals," Acta Materialia, vol. 55, pp. 4041–4065.
74. Y. Xiang, T. Y. Tsui and J. J. Vlassak (2006) "The mechanical properties of freestanding electroplated Cu thin films," Journal of Materials Research, vol. 21, pp. 1607–1618.
75. L. Lu, Y. Shen, X. Chen, L. Qian and K. Lu (2004) "Ultrahigh strength and high electrical conductivity in copper," Science, vol. 304, pp. 422–426.
76. O. Kraft, L. B. Freund, R. Phillips and E. Arzt (2002) "Dislocation plasticity in thin metal films," MRS Bulletin, vol. 27(1), pp. 30–37.
77. J. M. Jungk, W. M. Mook, M. J. Cordill, M. D. Chambers, W. W. Gerberich, D. F. Bahr, N. R. Moody and J. W. Hoehn (2004) "Length scale based hardening model for ultra-small volumes," Journal of Materials Research, vol. 19, pp. 2812–2821.
78. Y. Xiang and J. J. Vlassak (2005) "Bauschinger effect in thin metal films," Scripta Materialia, vol. 53, pp. 177–182.
79. C. J. Bayley, W. A. M. Brekelmans and M. G. D. Geers (2007) "A three-dimensional dislocation field crystal plasticity approach applied to miniaturized structures," Philosophical Magazine, vol. 87, pp. 1361–1378.
80. Y. Xiang and J. J. Vlassak (2006) "Bauschinger and size effects in thin-film plasticity," Acta Materialia, vol. 54, pp. 5449–5460.
81. M. Hommel, O. Kraft and E. Arzt (1999) "A new method to study cyclic deformation of thin films in tension and compression," Journal of Materials Research, vol. 14, pp. 2373–2376.
82. H. Huang and F. Spaepen (2000) "Tensile testing of free-standing Cu, Ag and Al thin films and Ag/Cu multilayers," Acta Materialia, vol. 48, pp. 3261–3269.
83. J. A. Ruud, D. Josell, F. Spaepen and A. L. Green (1993) "A new method for tensile testing of thin films," Journal of Materials Research, vol. 8, pp. 112–117.
84. J. A. Rogers, Z. Bao, K. Baldwin, A. Dodabalapur, B. Crone, V. R. Raju, V. Kuck, H. Katz, K. Amundson, J. Ewing and P. Drzaic (2001) "Paper-like electronic displays: large-area

rubber-stamped plastic sheets of electronics and microencapsulated electrophoretic inks," Proceedings of the National Academy of Sciences, vol. 98, pp. 4835–4840.

85. S. R. Forrest (2004) "The path to ubiquitous and low-cost organic electronic appliances on plastic," Nature, vol. 428, pp. 911–918.

86. R. H. Reuss, D. G. Hopper and J.-G. Park (2006) "Macroelectronics," MRS Bulletin, vol. 31(6), pp. 447–450.

87. T. Li, Z. Suo, S. P. Lacour and S. Wagner (2005) "Compliant thin film patterns of stiff materials as platforms for stretchable electronics," Journal of Materials Research, vol. 20, pp. 3274–3277.

88. F. Faupel, C. H. Yang, S. T. Chen and P. S. Ho (1989) "Adhesion and deformation of metal/polyimide layered structures," Journal of Applied Physics, vol. 65, pp. 1911–1917.

89. Y. S. Kang and P. S. Ho (1997) "Thickness dependent mechanical behavior of submicron aluminum films," Journal of Electronic Materials, vol. 26, pp. 805–813.

90. F. Macionczyk and W. Bruckner (1999) "Tensile testing of AlCu thin films on polyimide foils," Journal of Applied Physics, vol. 86, pp. 4922–4929.

91. M. Hommel and O. Kraft (2001) "Deformation behavior of thin copper films on deformable substrates," Acta Materialia, vol. 49, pp. 3935–3947.

92. D. Y. W. Yu and F. Spaepen (2004) "The yield strength of thin copper films on Kapton," Journal of Applied Physics, vol. 95, pp. 2991–2997.

93. J. Bohm, P. Gruber, R. Spolenak, A. Stierle, A. Wanner and E. Arzt, (2004) "Tensile testing of ultrathin polycrystalline films: a synchrotron-based technique," Review of Scientific Instruments, vol. 75, pp. 1110–1119.

94. G. P. Zhang, C. A. Volkert, R. Schwaiger, P. Wellner, E. Arzt and O. Kraft (2006) "Length-scale-controlled fatigue mechanisms in thin copper films," Acta Materialia, vol. 54, pp. 3127–3139.

95. S. H. Oh, M. Legros, D. Kiener, P. Gruber and G. Dehm (2007) "In situ TEM straining of single crystal Au films on polyimide: change of deformation mechanisms at the nanoscale," Acta Materialia, vol. 55, pp. 5558–5571.

96. J. N. Florando and W. D. Nix (2005) "A microbeam bending method for studying stress-strain relations for metal thin films on silicon substrates," Journal of the Mechanics and Physics of Solids, vol. 53, pp. 619–638.

97. Y.-L. Shen and S. Suresh (1995) "Elastoplastic deformation of multilayered materials during thermal cycling," Journal of Materials Research, vol. 10, pp. 1200–1215.

98. M. Ataka, A. Omodaka, N. Takeshima and H. Fujita (1993) "Fabrication and operation of polyimide bimorph actuators for a ciliary motion system," Journal of Microelectromechanical Systems, vol. 2, pp. 146–150.

99. R. B. Darling, J. W. Suh and T. A. Kovacs (1998) "Ciliary microactuator array for scanning electron microscope positioning stage," Journal of Vacuum Science and Technology A, vol. 16, pp. 1998–2002.

100. M. H. Mohebbi, M. L. Terry, K. F. Bohringer, G. T. Kovacs and J. W. Suh (2001) "Omnidirectional walking microrobot realized by thermal microactuator arrays," in Proceedings of the 2001 ASME International Mechanical Engineering Congress and Exposition, American Society of Mechanical Engineers, New York, paper no. 23824.

101. E. Arzt, G. Dehm, P. Gumbsch, O. Kraft and D. Weiss (2001) "Interface controlled plasticity in metals: dispersion hardening and thin film deformation," Progress in Materials Science, vol. 46, pp. 283–307.

102. H. J. Frost and M. F. Ashby (1982) Deformation mechanism maps, Pergamon Press, Oxford.

103. Y.-L. Shen and S. Suresh (1996) "Steady-state creep of metal-ceramic multilayered materials," Acta Materialia, vol. 44, pp. 1337–1348.

104. G. B. Gibbs (1966) "Diffusion creep of a thin foil," Philosophical Magazine, vol. 13, pp. 589–593.

105. M. D. Thouless (1993) "Effect of surface diffusion on the creep of thin films and sintered arrays of particles," Acta Metallurgica et. Materialia, vol. 41, pp. 1057–1064.

106. H. Gao, L. Zhang, W. D. Nix, C. V. Thompson and E. Arzt (1999) "Crack-like grain boundary diffusion wedges in thin metal films," Acta Materialia, vol. 47, pp. 2865–2878.

107. A. Gangulee (1974) "Strain-relaxation in thin films on substrates," Acta Metallurgica, vol. 22, pp. 177–183.
108. M. Murakami (1978) "Thermal strain in lead thin films II: strain relaxation mechanisms," Thin Solid Films, vol. 55, pp. 101–111.
109. D. Josell, T. P. Weihs and H. Gao (2002) "Diffusional creep: stresses and strain rates in thin films and multilayers," MRS Bulletin, vol. 27(1), pp. 39–44.
110. Y. H. Zhang and M. L. Dunn (2004) "Geometric and material nonlinearity during the deformation of micron-scale thin-film bilayers subject to thermal loading," Journal of the Mechanics and Physics of Solids, vol. 52, pp. 2101–2126.
111. A. Fisher-Cripps (2002) Nanoindentation, Springer, New York.
112. W. C. Oliver and G. M. Pharr (1992) "An improved technique for determining hardness and elastic modulus using load and displacement sensing indentation experiments," Journal of Materials Research, vol. 7, pp. 1564–1538.
113. W. C. Oliver and G. M. Pharr (2004) "Measurement of hardness and elastic modulus by instrumented indentation: Advances in understanding and refinements to methodology," Journal of Materials Research, vol. 19, pp. 3–20.
114. A. Gouldstone, N. Challacoop, M. Dao, J. Ki, A. M. Minor and Y.-L. Shen (2007) "Indentation across size scales and disciplines: recent developments in experimentation and modeling," Acta Materialia, vol. 55, pp. 4015–4039.
115. C. A. Schuh (2006) "Nanoindentation studies of materials," Materials Today, vol. 9(5), pp. 32–40.

Chapter 4
Patterned Films in Micro-devices

In this chapter attention is directed to patterned thin-film structures, where the film material exists as individual lines of various cross-section geometries. Contrary to the case of continuous films in Chap. 3, the deformation field in the line structure is dominated by the edge effect. In addition, the film segment may be entirely surrounded by one or more different materials so a severely confined condition is in place. The most representative example is the metal interconnects in modern integrated circuits. The interconnect structure is composed of several layers of Cu or Al lines embedded within the dielectric material (traditionally silica glass based, SiO_x) on top of the Si substrates. They serve as the connection between the functional elements (transistors) and between the transistors and the outside packaging structure. A schematic illustrating a two-level interconnect structure is shown in Fig. 4.1 (see also Fig. 1.4).

The structural integrity of interconnect materials has long been a concern in microelectronics. Possible failure modes include stress-induced voiding, electro-migration failure, interfacial delamination, and dielectric cracking, all of which are directly related to mechanical stresses. Constrained deformation in the metal lines plays an essential role in many aspects of interconnect failure. The thermal expansion mismatch between different materials is the most obvious source of stresses and deformation because the entire structure experiences thermal histories during fabrication and service. Stresses can also be generated during film deposition. The chemical and mechanical processes involved in patterning and planarizing, the non-uniform transport of atoms in the conducting lines (electromigration), the chemical reactions, and the absorption of foreign substances (such as water) all contribute to the generation of internal stresses. Furthermore, the rapid advancement of technology and miniaturization constantly presents new challenges. For instance, transistor scaling for meeting the performance target necessitates interconnect scaling, which renders the interconnect propagation delays a significant problem in microprocessors [1]. Replacing Al interconnects by Cu has helped with reducing the propagation delay by way of lowering the electric resistance. Further improvements can be made by lowering the capacitance of the dielectric

Y.-L. Shen, *Constrained Deformation of Materials: Devices, Heterogeneous Structures and Thermo-Mechanical Modeling*, DOI 10.1007/978-1-4419-6312-3_4,
© Springer Science+Business Media, LLC 2010

Fig. 4.1 Schematic
of a generic interconnect
structure consisting of two
levels of metal lines inside
the dielectric material.
Different levels of metal
lines are connected by vias

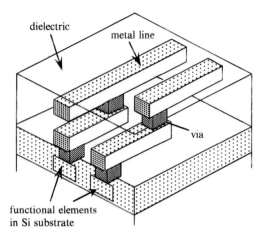

material used between the metal lines. The consequence, however, is that the
dielectric materials become increasingly weak mechanically, due partly to the
porosity introduced for reducing the dielectric constant. Thermo-mechanical
integrity is thus critical to the success of these new technologies. Background
information and case histories about the above mentioned concerns can be found
in many review articles and books [2–26]. The discussion in the following sections
is centered around stresses and constrained deformation in metal interconnects,
as well as their implications in device reliability.

4.1 Basic Consideration

As described in Chap. 3 a commonly used technique for measuring stress in a
continuous thin film attached to a thick substrate is the water curvature method.
The biaxial stress in the film causes the substrate to deform elastically through
bending. By measuring the curvature the stress in the film can be uniquely
determined. This technique has been extended to estimating stress in the film
patterned into a series of parallel lines [27–30]. A schematic of the patterned
line series is shown in Fig. 4.2. Applying the curvature method to passivated
(encapsulated) line structures has also been reported [31–34]. The X-ray
diffraction [35–54], Raman spectroscopy [55, 56], and micromachine-based
[57] techniques have also been developed and employed in measuring stresses
in specially designed metal line structures. On the theoretical side, there have
been analytical or semi-analytical formulations of thermal stresses in structures
composed of clamped elastic plates [58, 59] or a series of parallel lines on a
substrate [27, 60–70]. Computational finite element modeling can be employed
to account for complex geometry and material properties when predicting non-
uniform stress and strain distributions.

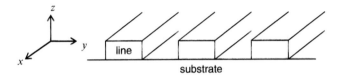

Fig. 4.2 Schematic of a series of parallel lines on top of a substrate

4.1.1 Elastic Lines on a Thick Substrate

4.1.1.1 Model Setup

We now consider an elastic structure of the same type as in Fig. 4.2, with the lines and substrate being SiO$_2$ (thermal oxide) and Si, respectively. The thermal mismatch induced and processing (film-etching) induced deformations can be examined within a unified modeling framework [29]. The model is similar to that considered in Fig. 3.3a, with the thin-film part being replaced by the film or line cross section as shown in Fig. 4.3a. The z-axis is now the reference mirror symmetry axis along which no displacement is allowed in the y-direction during deformation. The right boundary ($y=p/2$ before deformation) is also a mirror symmetry axis; in general it can deviate from the initial vertical position but is forced to remain a straight line during deformation. All other boundaries are not constrained. The upper SiO$_2$ region of the structure in Fig. 4.3a contains two parts. The case of a continuous thin film is in place when both parts are present. An internal biaxial stress state can be built into the model. When simulating etching of the film into parallel lines, elements in the right part ($w/2 < y < p/2$ and $0 < z < h$) are removed while the static equilibrium condition is maintained. When simulating thermal loading of the line structure, only the left part ($0 < y < w/2$ and $0 < z < h$) is included and a spatially uniform temperature change is imposed. Note the symbols w and p represent the line width and pitch (periodicity) of the line array, respectively, and h and h_s are the thicknesses of the line and substrate, respectively.

In addition to obtaining the stress and deformation fields in the structure, the evolution of overall curvature can also be directly calculated from the current model. The curvature in the y-direction is determined from the relative positions of the two sides $y=0$ and $y=p/2$, i.e.,

$$\kappa_y = \frac{2}{h_s \cdot p} \cdot \left\{ \left[u_y \right]_{y=p/2, z=-h_s} - \left[u_y \right]_{y=p/2, z=0} \right\}, \tag{4.1}$$

where u_y is the displacement along the y-direction. For obtaining the curvature along the x-direction, a generalized plane strain formulation, which is an extension of the plane strain framework (with the yz-plane being the plane of deformation) is used. This is achieved by superimposing a longitudinal strain field, ε_{xx}, on the

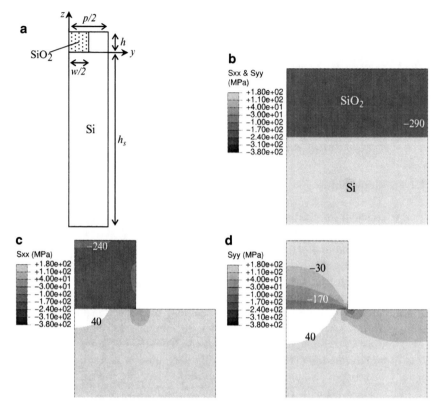

Fig. 4.3 (a) Schematic of the SiO$_2$ line/Si substrate model used in the finite element analysis. The x-axis is defined to be along the SiO$_2$ line direction (out-of-paper). (b) Contour plot of σ_{xx} and σ_{yy} (an equi-biaxial stress state) before etching of the SiO$_2$ film into lines. The stress value in SiO$_2$ is −290 MPa. (c) Contour plot of σ_{xx} after etching. (d) Contour plot of σ_{yy} after etching

plane strain state. To properly simulate the actual response of the parallel lines on the substrate, the strain field ε_{xx} is constrained to induce a constant rotation about the y-axis:

$$\frac{\partial^2 u_z}{\partial x^2} = C, \tag{4.2}$$

where u_z is the displacement along the z-direction and C is a constant directly determined by the analysis. The 3D effect can thus be adequately described by the present model. Equation (4.2) directly gives the wafer curvature in the x (line)-direction,

$$\kappa_x = \frac{\partial^2 u_z}{\partial x^2} = C. \tag{4.3}$$

The generalized plane strain model is thus capable of yielding more realistic results than the strict plane strain formulation.

In the simulation a thermal cooling is first performed on the model with a continuous SiO_2 thin film, to introduce an equi-biaxial stress field of value -290 MPa in the film. This value is based on an experimental measurement of a thermally grown SiO_2 film on a Si substrate. Figure 4.3b shows a contour plot of σ_{xx} (and σ_{yy}) in SiO_2 and the top portion of Si at this stage. Subsequently, etching of the film into lines with an aspect ratio (h/w) of 0.537 and pitch-to-width ratio (p/w) of 2.286 is simulated. The resulting stress fields of σ_{xx} and σ_{yy} are shown in Fig. 4.3c and d, respectively. It can be seen that the stress field becomes non-uniform upon etching. The magnitude of σ_{xx} in the SiO_2 line is moderately reduced (Fig. 4.3c). Because the etched line maintains a continuous form along this (x) direction, the stress is thus not relieved to a great extent. The magnitude of σ_{yy} in SiO_2, however, is significantly reduced. High stresses exist only in a very narrow area adjacent to the substrate. This is due to the existence of vertical side walls, on which the traction-free condition has to be satisfied.

If a temperature change is directly imposed on the patterned line structure from a stress-free state, thermal expansion induced deformation will occur. The resulting stress and deformation fields, however, are equivalent to those obtained from the etching process, provided that the temperature change imposed on the line/substrate structure is the same as that on the film/substrate structure prior to etching. This is because linear elasticity produces a thermo-mechanical response which is independent of the loading path. (In other words, the sequence of cooling and then etching leads to the same outcome as the sequence of etching and then cooling, when the material exhibits a linear elastic behavior.) Therefore, the results presented in this section are valid for both the etching and thermal loading processes.

4.1.1.2 Parametric Analysis

A parametric analysis can be conducted taking into account a wide range of line geometries. Figure 4.4 shows the variation of line stress, averaged over the cross section of the line, with the line aspect ratio from the numerical analysis. Here the aspect ratio is defined to be h/w. In the limiting case of zero aspect ratio, an equi-biaxial stress state results, i.e., $\sigma_{xx} = \sigma_{yy} = \sigma_{film}$ where the subscript "film" represents the case of a continuous film. As the aspect ratio increases, σ_{yy} decreases precipitously due to the strong influence of the vertical side walls of lines, as expected. The extent of such a decrease for σ_{xx}, however, is much less pronounced. At the other extreme when the aspect ratio approaches infinity (a very tall and narrow line structure),

$$\sigma_{xx} = E_{SiO_2} \varepsilon_{xx}. \qquad (4.4)$$

Furthermore,

$$\sigma_{film} = \frac{E_{SiO_2}}{1 - v_{SiO_2}} \varepsilon_{film}. \qquad (4.5)$$

Here, as before, E and v are Young's modulus and Poisson's ratio, respectively. Combining (4.4) and (4.5) and taking into account that the strain $\varepsilon_{xx} = \varepsilon_{film}$, one obtains

$$\frac{\sigma_{xx}}{\sigma_{film}} = 1 - v_{SiO_2} = 0.84. \qquad (4.6)$$

as the line aspect ratio approaches infinity. (Note in the present simulation the Poisson's ratio of SiO_2 is taken to be 0.16 [29].) It is evident that the numerical prediction (Fig. 4.4) shows such a relation. Another important result from the same modeling study is that the interspacing between the parallel lines plays a very insignificant role in affecting the average stress in lines.

Attention is now turned to the wafer curvature. Figure 4.5 shows a comparison between modeling and experiments, where the ratio of curvatures along the two directions (κ_y/κ_x) is plotted against the line aspect ratio. The experimental results include both the etching and thermal loading processes conducted independently. The simulated curvature evolution showed agreement with the experimental curvature results. In the limiting case of a continuous film, the aspect ratio is zero and hence the curvatures are equal in both directions. Figure 4.5 shows that a negative curvature ratio (i.e., a "potato chip" shaped wafer) exists for an aspect ratio greater than about 0.4. By considering the plate bending problem in elasticity [71], the average stresses in the patterned film can be related to the curvature as

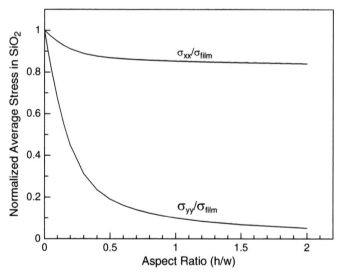

Fig. 4.4 Numerically predicted average stresses in SiO_2 lines, σ_{xx} and σ_{yy}, normalized by the corresponding film stress, σ_{film}, as a function of line aspect ratio [29]. The modeling uses a pitch-to-width (p/w) ratio of 1.5; different p/w ratios give essentially identical curves

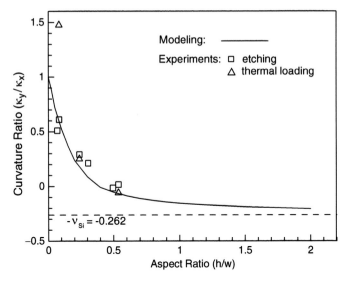

Fig. 4.5 Numerically predicted and experimentally measured curvature ratios as a function of line aspect ratio [29]. A *dashed line* corresponding to $\kappa_y/\kappa_x = -v_{Si}$ is also included. The modeling uses a pitch-to-width (p/w) ratio of 2.3; different p/w ratios give essentially identical curves

$$\sigma_{xx} = \frac{E_{Si}h_{Si}^2}{6(1-v_{Si}^2)h_{SiO_2}}\left(\kappa_x + v_{Si}\kappa_y\right), \tag{4.7}$$

$$\sigma_{yy} = \frac{E_{Si}h_{Si}^2}{6(1-v_{Si}^2)h_{SiO_2}}\left(\kappa_y + v_{Si}\kappa_x\right), \tag{4.8}$$

where E, v, h and κ stand for Young's modulus, Poisson's ratio, thickness (height) of the layer and curvature, respectively. The assignment of coordination axes follows that in Fig. 4.3. When the aspect ratio approaches infinity, σ_{yy} vanishes and therefore, from (4.8),

$$\frac{\kappa_y}{\kappa_x} \to -v_{Si}, \tag{4.9}$$

which is indeed observed in Fig. 4.5 at large aspect ratios. (Note in the present simulation the Poisson's ratio of Si(111) wafer is taken to be 0.262.) It is also worth mentioning that the modeling curve is based on a pitch-to-width ratio (p/w) of 2.3; using other p/w ratios results in essentially identical modeling results.

The sign change of the curvature ratio at h/w about 0.4 is caused by the reversal of curvature in the direction perpendicular to the lines (κ_y). Its physical nature is schematically illustrated in Fig. 4.6. In this illustration the bending configurations are based on the response of thermal heating rather than cooling of the structure. In Fig. 4.6a, the aspect ratio of the lines is low so the situation is qualitatively

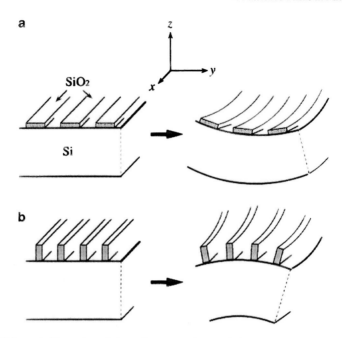

Fig. 4.6 Schematic illustration of the wafer shape when the SiO$_2$ line aspect ratio is (**a**) very low and (**b**) very high. In this example an increase in temperature from the initial state of flat shape is considered

similar to that of a continuous film. Bending along both the x and y axes occurs in the same direction (both concave upward). When the line aspect ratio is very high as in Fig. 4.6b, the average stress σ_{xx} in lines remains high but σ_{yy} is very low (Fig. 4.4). The constraint imposed on the Si substrate along the y-direction is essentially lost. As a consequence, the wafer remains concave upward in x, but the Poisson's effect forces an expansion and contraction in, respectively, the upper and lower portions of Si so it becomes concave downward in y. A "potato chip" shape is thus formed.

4.1.2 Elastic-Plastic Lines on a Thick Substrate

A similar type of procedure can be applied to the case of patterned thin films having the capability of plastic yielding. We now consider the example of a series of parallel aluminum lines on a silicon substrate. Stresses in the Al lines, resulting from etching of a continuous film bearing an initial stress, can be directly calculated. The variation of stresses σ_{xx} and σ_{yy} with line aspect ratio follows the same trend as in Fig. 4.4. Although plastic deformation of the metal is accounted for in the model, the line stress is largely conformable to the elastic analysis similar to (4.6); in this case

$$\frac{\sigma_{xx}}{\sigma_{film}} = 1 - v_{Al} = 0.67 \qquad (4.10)$$

when the aspect ratio approaches infinity. This similarity arises from the fact that etching causes elastic unloading in nearly the entire volume of the line. The variation of stress during etching follows the same pattern as the elastic case except in the region very close to the corner of the line-substrate interface where a stress concentration exists [72]. Although (4.10) is true to the case of a very large line aspect ratio, its accuracy is in fact reasonably good for any line aspect ratio greater than about 1.0. Therefore, for estimating the longitudinal line stress in an *unpassivated* test structure (such as those commonly used in electromigration testing), (4.10) serves as a good rule-of-thumb for narrow lines. The detailed stress distribution in the cases of Al and gold (Au) lines of various aspect ratios can be found in references [72–75].

In the following we consider an example of stress history in unpassivated Al lines during temperature cycles. The finite element model is the same as in Fig. 4.3a, except that the line material is now Al. There is no etching process considered, with the beginning structure already consisting of patterned lines. The Al material is assumed to be elastic-perfectly plastic with a temperature dependent yield strength following the experimental measurement of the 1-μm thick continuous film (Fig. 3.4). The material properties used in the present simulation, as well as in most of the modeling analyses presented in Chap. 4, are listed in Table 4.1. The stress-free reference temperature is taken to be 450°C. Following the initial cooling to 20°C, the evolution of stresses during a full thermal cycle of 20°C → 450°C → 20°C is monitored.

Figure 4.7a shows the stresses σ_{xx} and σ_{yy}, in the Al lines having an aspect ratio $h/w = 1$, as a function of temperature during the temperature cycle. It can be seen that the lines are able to support a much higher stress in the longitudinal direction

Table 4.1 Material properties used in most of the numerical analyses considered in Chap. 4

	Al	Cu	SiO$_2$	Si	TaN	SiN$_x$	BCB	SiCOH
E (GPa) at 20°C	69	110	71.4	130	200	221	2.5	6.0
400°C	55	103.1	71.4	130	200	221	0.3[a]	6.0
v at 20°C	0.33	0.3	0.16	0.28	0.3	0.27	0.34	0.3
400°C	0.33	0.3	0.16	0.28	0.3	0.27	0.34	0.3
α (10^{-6}/K) at 20°C	23.6	17.0	0.52	3.1	4.7	3.2	63.6	20
400°C	29.9	19.6	0.73	4.7	4.7	3.2	63.6	20
σ_y (MPa) at 20°C	200	Sect. 3.2	–	–	–	–	–	–
400°C	67.4	Sect. 3.2	–	–	–	–	–	–

Unless otherwise stated, a linear variation of properties with temperature defined by the indicated temperatures is assumed. Some material properties are shown as independent of temperature, because the temperature dependence of these properties is either unclear or the incorporation of temperature variation will only affect the modeling result very slightly

E: Young's modulus, v: Poisson's ratio, α: coefficient of thermal expansion, σ_y: yield strength
[a]Linear variation only between 20 and 180°C [76]; a constant value (0.3 GPa) above 180°C

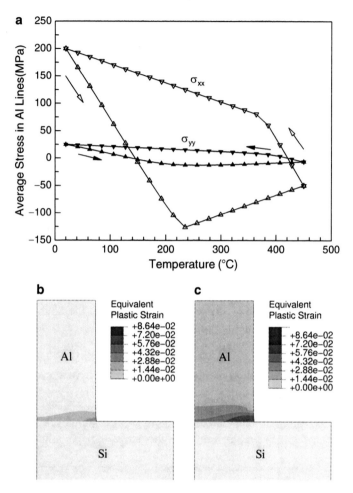

Fig. 4.7 (a) Variation of the stress components σ_{xx} and σ_{yy} as a function of temperature during a thermal cycle between 20 and 450°C obtained from the finite element modeling. The stresses are averaged over the cross section of the Al lines with an aspect ratio of unity. The definition of the coordinate system is the same as in the previous figures: x is parallel to the line and y is across to the line. (b) and (c) Contour plots of equivalent plastic strain after initial cooling (in (b)) and after a subsequent full thermal cycle (in (c))

(x) than in the transverse direction (y), as expected. The fundamental features of the stress–temperature curves, however, are similar. At the beginning of the heating and cooling phases, the Al lines largely behave elastically. Gross plastic deformation commences at temperatures about 230 and 380°C during heating and cooling, respectively. One notable observation is that the σ_{xx} response in Fig. 4.7a is quantitatively very similar to the biaxial stress in a continuous film (Fig. 3.4). The magnitude of σ_{yy} will depend strongly on the line aspect ratio: it increases as

h/w decreases. In the limiting case of $h/w \approx 0$, the result of a continuous film will be recovered.

Figure 4.7b and c show the contour plots of equivalent plastic strain after the initial cooling and after a subsequent thermal cycle, respectively, in the Al lines with an aspect ratio of unity. Concentration of plastic deformation in the metal occurs mainly along the interface with the Si substrate, especially near the corner region. However, the entire Al line has yielded. By comparing the contour shades in parts (b) and (c), it is seen that a significant amount of plasticity has accumulated over one cycle between 20 and 450°C.

One legitimate concern regarding the modeling above is that the metal line is assumed to be isotropic with the same yield behavior as its continuous-film counterpart. Is this a reliable approximation to represent the actual material? The effect of anisotropic yield properties can be studied, within the current continuum framework, by treating the line as transversely isotropic with regard to the yield strength. This is a simple way to simulate possible directionality of yield behavior arising from the non-equiaxed grain structure within the patterned lines, without recourse to more elaborate modeling techniques. The yield strength in the x-direction can be taken as that of the continuous film (σ_{film}); various values can be assigned as the yield strength in the y- and z-directions. It has been shown that setting the transverse yield strength as $0.5\sigma_{film}$ and $1.5\sigma_{film}$ results in only small differences in both σ_{xx} and σ_{yy} throughout the temperature range [72]. As a consequence, using the plastic properties of the continuous film in modeling of the line structure is generally a safe approach.

4.2 Passivated Single-Level Lines

In actual devices metal interconnects are almost always encapsulated by the dielectric and/or passivation layer. The confinement of the metal has prompted some researchers to presume a pure elastic state in analyzing stresses, arguing that a near hydrostatic stress field caused by thermal mismatch actively suppresses plastic yielding. In reality, the metal is far from being confined in a rigid box, as illustrated in an early finite element analysis [77]. The rectangular cross-section geometry, a dominant longitudinal dimension of the line, and the non-infinite and non-rigid nature of the surrounding material all contribute to the generation of deviatoric stress components, leading to plastic deformation in the embedded metal line. Finite element modeling on Al interconnects has shown that, if the metal line is treated as purely elastic, significantly higher stresses would be obtained [74]. Incorporating plasticity eliminates the large stress gradient caused by the elastic assumption, and results in more accurate stress magnitudes within the metal line. In the following we discuss the evolution of stress and deformation fields in Al and Cu interconnect lines separately. Attention is devoted to structures with a series of parallel, single-level lines.

4.2.1   Aluminum Interconnects

We first consider a simplified model of Al line embedded within the SiO_2 dielectric, as shown in Fig. 4.8a. As before the longitudinal (x) direction is perpendicular to the paper, and the calculation follows the generalized plane strain formulation. The Si substrate is excluded from the present model (note in actual devices Al lines are not in direct contact with Si.) The passivation geometry and properties are

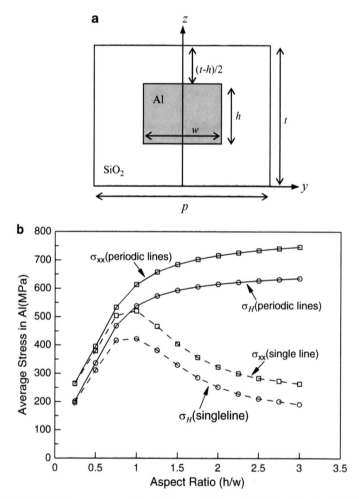

Fig. 4.8 (a) The Al interconnect model, with the metal line embedded within the SiO_2 dielectric. (b) Predicted average longitudinal stress σ_{xx} and hydrostatic stress σ_H in Al as a function of line aspect ratio (h/w), for the case of $p/w = 2$. Both the results of periodically arrayed Al lines (*solid curves*) and a single Al line (*dashed curves*) are included. In the case of "periodic lines" the side boundaries of SiO_2 are constrained to remain straight and vertical; for "single line" the side boundaries are not constrained

known to affect the line stress [74, 78]; here a planarized passivation is assumed and the total height of SiO_2 is taken to be twice that of Al [79]. The top boundary is free to move during deformation, while the bottom boundary is assumed to be rigidly clamped to a substrate so no displacement is allowed in both the y- and z-directions. The side boundary conditions are discussed in the following paragraph. Note that due to symmetry, the actual simulation can use only one half of the structure shown in Fig. 4.8a. The material properties used in the modeling are listed in Table 4.1.

Figure 4.8b shows the modeled stresses in the Al line as a function of the line aspect ratio h/w. The stresses shown are volume averaged values of the longitudinal component, σ_{xx}, and hydrostatic component, σ_H, after a thermal history of cooling from 400 to 20°C, where 400°C is taken to be the initial stress-free state (presumably the passivation deposition temperature or film annealing temperature). In addition, the solid curves in Fig. 4.8b, denoted "periodic lines," pertain to a pitch-to-width ratio (p/w) of 2, meaning that there is an infinite array of lines and the line spacing is equal to the line width. Under this circumstance the side boundaries in Fig. 4.8a are constrained to remain straight and vertical during deformation. It can be seen that the stresses increase monotonically with increasing line aspect ratio from below one to well above one. For an aspect ratio greater than about 1.5, the increasing rates of both stress components become small, however, the stress magnitudes are still increasing. Also included in Fig. 4.8b are the results for the "single line" case, representing an embedded isolated line (i.e., the side boundaries of SiO_2 are free of constraint so the interaction from its neighboring lines in the periodic case is eliminated). It can be seen that for an aspect ratio less than about 0.5, both the longitudinal and hydrostatic stress components in the single Al line show similar magnitudes as in the case of the periodic Al lines. The stresses in the single line increase with the line aspect ratio and attain maximum values when the aspect ratio is unity, beyond which they decrease drastically. At high aspect ratios, the stresses are much lower than their periodically arranged counterparts. This is simply because the side boundaries of the single-line structure are unconstrained; a greater line height causes stress relief in a greater portion of the line.

The results in Fig. 4.8 bring about an important issue: one must consider the interaction effect due to neighboring lines. There have been numerical and analytical studies claiming that the maximum values of normal stress components in metal interconnects occur when the line aspect ratio is equal to one. These calculations, however, were based on idealized geometries of either the single-line configuration [74] or an isolated Al line embedded within an infinitely large matrix [60, 80]. Figure 4.8b clearly illustrates that the maximum stresses in lines do not exist when the line width is equal to line height. The importance of using proper boundary conditions in modeling interconnect deformation cannot be overemphasized, especially when the lines are closely spaced as commonly encountered. There have been other calculations [33, 44] based on the unit-segment model for analyzing or interpreting experimental results. In these studies an isolated Al line, instead of periodically arrayed lines used in their experiments, was assumed. Again, care should be taken when doing so because, for the same reason just described, large errors can be induced for common line geometries. Further analyses, including the effects of line spacing, have also been reported [72, 79].

There are other material and geometric factors that can influence the Al line stress. Finite element modeling on the basis of time-dependent creep properties of Al has been reported [81]. However, since the thermal cycling response of Al (and Cu) films is found to be insensitive to the heating/cooling rate above the range of 1–5°C per minute [82, 83], using the rate-independent elastic-plastic response should be a satisfactory phenomenological approach in normal practices. As of geometric factors, a case in point is the thickness of SiO_2. A thinner dielectric imposes less constraint on the embedded metal, and therefore results in lower stresses. Even the thickness of SiO_2 *below* the metal plays a role [84]. A thicker SiO_2 layer below Al tends to reduce the stress in the metal line. Another issue of concern is the thin barrier layers cladding the Al lines, such as titanium nitride (TiN) serving to improve processing and reliability conditions, in actual devices. It has been found that the tensile stress in Al is enhanced, especially in high-aspect-ratio lines, if the thin TiN layers are incorporated in the numerical model [85]. This is due to the mechanically stiff nature of the refractory film.

4.2.2   Copper Interconnects

The material layout in the case of Cu interconnects in modern microelectronic devices is more complicated. Figure 4.9a shows a model scheme for a series of infinite parallel Cu lines. The Cu lines in this model are encased by the thin silicon nitride (SiN_x) and tantalum nitride (TaN) diffusion barrier layers. Due to symmetry only one half of the unit segment is shown. Here the thick Si substrate is included in the analysis. The boundary conditions are identical to those considered in Sect. 4.1: along the left boundary (z-axis) no displacement is allowed in y; the right boundary is allowed to deviate from being vertical but has to remain straight during deformation. The input material parameters used in the simulation are listed in Table 4.1, with the kinematic hardening feature of the elastic-plastic Cu thin film accounted for (see Sect. 3.2 for details). For comparison purpose a purely elastic Cu, without any plastic deformation allowed in the model, is also considered. The initial stress-free temperature is taken to be 350°C.

4.2.2.1   Elastic Versus Elastic-Plastic Copper Lines

Figures 4.9b and c show the contour plots of hydrostatic stress in the Cu line and its vicinity, obtained from modeling using, respectively, the *elastic-plastic* model and *purely elastic* model for Cu, after cooling from 350 to 20°C. It is seen that high triaxial tension exists in Cu, and the stress field is non-uniform. Compared with the purely elastic Cu, the hydrostatic stresses in the elastic-plastic Cu are generally smaller, but the difference is not significant. Figure 4.9d shows the contours of

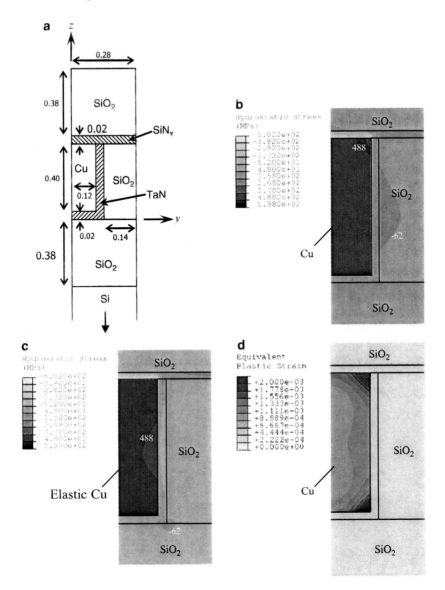

Fig. 4.9 (a) 2D Cu interconnect model (unit: μm). Only one half of the unit structure is shown, with the z-axis being the symmetry axis. (b) and (c) Contours of hydrostatic stress in Cu and its vicinity after cooling using elastic-plastic Cu (in [b]) and purely elastic Cu (in (c)). (d) Contours of equivalent plastic strain using elastic-plastic Cu

equivalent plastic strain in the copper line after cooling, for the case of elastic-plastic Cu. Plastic deformation occurs in the metal, especially in the corner regions and near the interfaces with the barrier layers. Although the numerical analysis is continuum-based, this observation nevertheless implies, from the materials science

point of view, the concentration of crystal defects, namely dislocations and
vacancies, near these locations. Subsequent void nucleation and electromigration
damages can be expected to preferentially occur in the regions. Experiments have
shown that the interface between Cu and the adjacent barrier layers is an easy
diffusion path and is particular prone to failure initiation [22, 25, 86–90]. The adhesion quality of the interface in actual devices plays a very important role. The present modeling also suggests that, even with a perfectly bonded interface, the
thermo-mechanical field in the interconnects favors local damage in the interface
region as well. If the Cu line is taken as purely elastic in the modeling, important
information will be lost [91].

The similarity of stress quantities in Fig. 4.9b and c implies that using experimental means (such as X-ray diffraction) to measure stresses in Cu lines may not
be able to capture the plastic nature of the metal. This is manifested by considering
the stress-temperature history resulting from the modeling. Figure 4.10 shows the
average hydrostatic stress σ_H and stress components σ_{xx}, σ_{yy} and σ_{zz} in the Cu line,
as a function of temperature when Cu is treated as a purely elastic (Fig. 4.10a) and
elastic-plastic (Fig. 4.10b) materials, during a stabilized thermal cycle (after initial
cooling) between 20 and 350°C. Note the stress-temperature representation in
Fig. 4.10 is in the same typical format for reporting interconnect stress measurement using the diffraction technique [51, 92, 93]. It can be seen that the stress
component in the longitudinal direction (σ_{xx}) is the highest among all normal components at room temperature. In Fig. 4.10a, the elastic nature of Cu results in the
same path during heating and cooling. When the plastic response of Cu is taken into
account, Fig. 4.10b, a hysteresis loop can be observed. It is noticed that the stress-
temperature loops in Fig. 4.10b are quite narrow, owing to the high strain hardening
rate of the Cu film. At a specific temperature, the difference in stress magnitudes
during heating and cooling is small and can be within experimental uncertainty. As
a consequence, an experimentally measured stress–temperature behavior like this
(such as those in Refs. [51, 92]) may lead to the belief that passivated Cu interconnect lines behave only elastically under the severe constraint, although plastic
deformation may well have occurred during the process.

4.2.2.2 Influence of Dielectric Material

One important aspect about deformation in Cu interconnects is the integration of
Cu with low-dielectric constant materials (low-k dielectrics) in advanced devices.
These dielectric materials (can be silica glass-based or polymer-based) typically
have a much greater CTE and much lower elastic modulus than the traditional SiO_2.
Therefore, their influence on the plastic deformation in Cu can be quite different.
Below we consider an example using essentially the same model setup as in
Fig. 4.9. The only difference in the present case is that the SiO_2 dielectric adjacent
to the side wall of the metal line in Fig. 4.9a is now replaced by the polymer-based
divinyl-siloxane-bis-benzo-cyclobutene low-k material (termed BCB henceforth).
Its thermo-mechanical properties are also listed in Table 4.1. Note the BCB properties

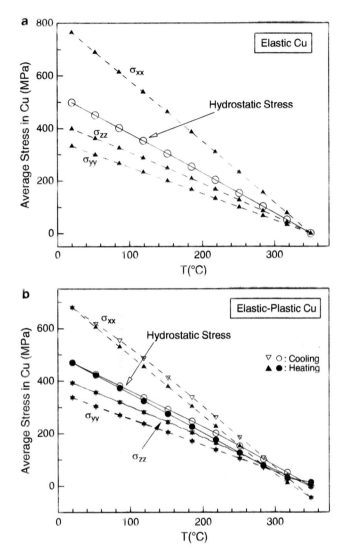

Fig. 4.10 Thermal stresses in the Cu line (averaged over the cross section) as a function of temperature obtained from the finite element modeling, assuming Cu as (**a**) purely elastic and (**b**) elastic-plastic. In both cases the thermal history involved is a full cycle from 20°C (after initial cooling) to 350°C and then back to 20°C [91]

are very similar to some other polymer-based low-k materials such as SiLK™ [94]. Note that the dielectric material above and below the metal structure is still SiO_2.

Figure 4.11a–d show the contour plots of stresses σ_{xx}, σ_{yy}, σ_{zz} and equivalent plastic strain, respectively, in the Cu line and its vicinity, after cooling down from 350 to 20°C. It can be seen that the stress state inside Cu deviates significantly from high triaxial tension. The longitudinal tensile stress is still quite high (Fig. 4.11a).

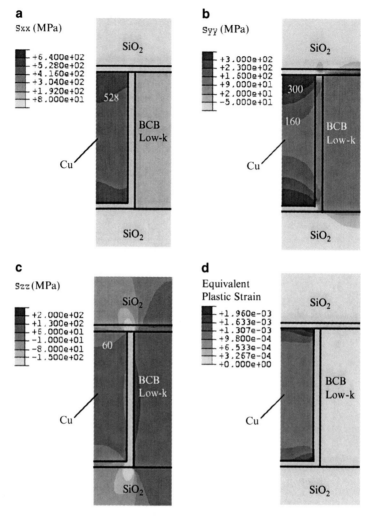

Fig. 4.11 Contour plots of (**a**) σ_{xx}, (**b**) σ_{yy}, (**c**) σ_{zz} and (**d**) equivalent plastic strain in Cu and its surrounding materials, after cooling from 350 to 20°C. The dielectric material adjacent to the side wall of the metal structure is BCB low-k polymer [95]

However, the magnitudes of σ_{yy} and σ_{zz} are significantly reduced. The effect on σ_{zz} is particularly distinct. Since the low-k dielectric has a CTE value greater than that of Cu, during cooling it actually tends to impose compressive σ_{zz} on Cu but is counterbalanced by the stiff barrier layer whose CTE is low. As a consequence, low tensile stresses result. Figure 4.11d shows that plastic deformation in Cu is localized in the top and bottom interface regions. For a quantitative comparison with the case of the all-SiO$_2$ dielectric, one can calculate the field quantities averaged over the cross section of the Cu line, and the results are listed in Table 4.2.

Table 4.2 Volume-averaged stress and plastic strain values in the Cu line, for the interconnect models considered in Fig. 4.9 (SiO$_2$ dielectric) and Fig. 4.11 (BCB low-k dielectric)

	σ_{xx} (MPa)	σ_{yy} (MPa)	σ_{zz} (MPa)	Hydrostatic stress (MPa)	Equivalent plastic strain
Cu/SiO$_2$	679	338	393	470	9.43×10^{-4}
Cu/BCB	517	223	81	274	1.22×10^{-3}

The cases of SiO$_2$ and BCB low-k dielectric correspond to those of Figs. 4.9 and 4.11, respectively. The very large difference in σ_{zz} is evident. Although there is a more than 40% decrease in hydrostatic tensile stress when SiO$_2$ beside the line is replaced with BCB, the plastic strain actually increases by about 30%. It is worth mentioning that, if the nitride barrier layers do not exist, the use of polymer-based low-k dielectric will result in even lower hydrostatic stress and higher equivalent plastic strain [95]. Further discussion on the dielectric-affected Cu line deformation will be given in Sect. 4.4.

4.3 Complex In-plane Geometries

In real devices the geometric layout can be more complex than the long and parallel lines considered thus far. One feature of practical significance is the sharp turns and branches, namely the commonly encountered L and T shaped metal lines. Schematics of two examples are shown in Fig. 4.12a and b. It is important to obtain a quantitative picture of the local stress and strain fields in the angled region, which provides indications of voiding propensity and serves as the initial condition for modeling other reliability related phenomena such as electromigration stress buildup. In the following we consider 3D finite element modeling of thermal stresses for single-level Al lines.

The simplified L and T shaped interconnect structures used in the analysis are shown in Fig. 4.13a and b, respectively [96]. The width and height of the Al line are represented by w and h, respectively. There is a silicon oxide (SiO$_2$) layer of thickness $0.5h$ directly below the metal line. A planarized SiO$_2$ passivation layer, with a thickness of $0.5h$ above the metal, is included. The bottom plane ($z=0$) is assumed to be clamped on a rigid substrate. The boundary conditions are such that mirror symmetry with respect to the side planes is imposed. Thus, along the planes $x=0$ and $x=p$, no displacement is allowed in the x direction; along the planes $y=0$ and $y=p$, no displacement is allowed in the y direction. The top surface is not constrained. The line aspect ratio (h/w) considered here is 1.0. The side length p is taken to be $7w$. This choice facilitates the condition that the "legs" of the metal line are short enough to preserve computational efficiency, but are sufficiently long so the line stress at locations away from the angled region is close to that in a very long straight line, as obtained in Sect. 4.2.1. The initial stress-free temperature is taken as 400°C. The relevant material properties used in the simulation are listed in Table 4.1.

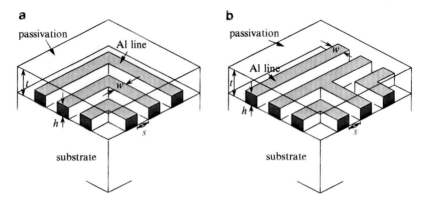

Fig. 4.12 (a) and (b) Examples of the more complex in-plane layouts of interconnect lines

Fig. 4.13 Schematics of the (a) *L* and (b) *T shaped Al lines* in the interconnect structure used in the modeling analysis

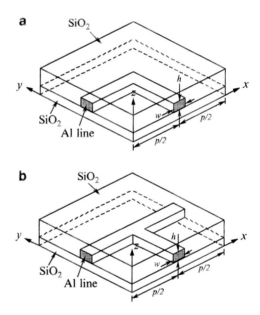

Figure 4.14a and b show the contour plots of hydrostatic stress and equivalent plastic strain, respectively, averaged over the line height, in the L shaped line, upon cooling from 400°C to room temperature. The corresponding contour plots for the T shaped line are shown in Fig. 4.14c and d. It can be seen from Fig. 4.14a that the convex corner of Al shows slightly higher tensile stresses than those in the straight segments. The stress concentration disappears in the T shaped line (Fig. 4.14c) because of the nonexistence of a convex corner. The concave corner of Al shows a moderate decrease in tensile stress. In the T shaped lines, the reduction of stress in

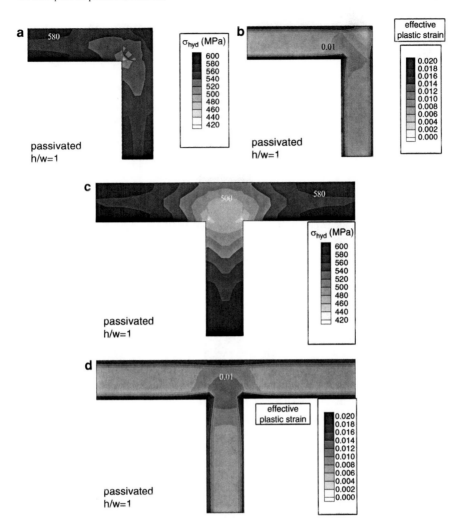

Fig. 4.14 Contour plots of (**a**) hydrostatic stress and (**b**) equivalent plastic strain in the passivated *L shaped Al line*; contour plots of (**c**) hydrostatic stress and (**d**) equivalent plastic strain in passivated *T shaped Al line*. The stress and strain magnitudes are averaged over the line height

the broad center region is evident. In both Fig. 4.14b and d, the plastic strain is seen to be enhanced around the concave corner but only in a very localized manner. Experimental micrographs have shown that stress induced voiding does not preferentially occur in the angled region [97]. Overall, there does not seem to be any significant *global* concentration of deformation field around the sharp turns and branches. It should be noted, however, that when an electric current exists, current crowding tends to occur near the concave corner so local electromigration damage may be enhanced.

Fig. 4.15 Contour plot
of hydrostatic stress in the
L shaped Al line with
a radius of curvature of 0.5 *w*
at the corners

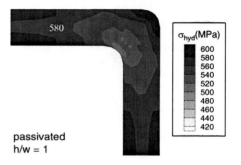

The information presented above is also useful in designing test structures for micro-scale stress measurement. Further analyses on the effect of line aspect ratio for both the passivated and unpassivated Al lines have also been conducted [96]. Another geometric factor of importance is the corner curvature. In actual interconnect structures the corners in L and T lines may not be very sharp as considered in the above example. The effects of corner curvature on the thermal stress evolution can be examined by modifying the local geometry in Fig. 4.13. Here we show an example of L shaped line with the radius of curvature at the corner taken to be 0.5w, a relatively large value. All the other modeling conditions remain unchanged. Figure 4.15 shows the contour plot of hydrostatic stress in the Al line. A comparison with Fig. 4.14a reveals that the overall stress patterns are similar. Only the stress distribution in the very localized region near the corner are moderately affected, even with this fairly large radius of curvature accounted for. If the radius of curvature is smaller than that considered here, the stress fields will be very close to those with sharp corners. The same was also found true in the cases of different line aspect ratios and of T shaped lines, with and without passivation. The stress field is thus dominated by the global line geometry, not the local corner curvature, of the interconnect. The above results were for Al lines. The effects of these geometrical features on Cu interconnects remain to be studied in a systematic way. This requires modeling taking into account the appropriate material properties and other barrier-layer and dielectric features.

4.4   Multilevel Structures

The foregoing discussions have concentrated on single-level metal lines. However, when addressing thermo-mechanical reliability the effects of the realistic multilevel structure should not be overlooked. For example, based on experimental damage observations, the possibility that the lower-level lines are under higher tensile stresses has been raised [2, 98]. It is clear that the current understanding of the problem is still inadequate.

4.4.1 Parallel Lines

Owing to the film deposition processes, metal lines at different levels experience different thermo-mechanical histories. In a finite element study taking into account the stress history during device manufacturing [99], it was illustrated that the final stress state in the metal lines at room temperature resulting from the extensive thermal excursions is essentially the same as that obtained from a single-step modeling of the final cooling process. This suggests that assuming a single cooling step from a stress-free temperature is a reasonable approach to calculate thermal stresses generated in multilevel metal lines. On this basis, a numerical study considered the evolution of thermal stresses in multilevel interconnects with up to four levels of Al lines with various cross-section arrangements and aspect ratios [85]. If all lines have the same aspect ratio, the stresses in the metal at different levels were found to vary only moderately with no clear trend. A staggered arrangement results in lower tensile stresses compared to the vertically aligned arrangement. There is no general propensity of increasing stresses from the upper-level to lower-level lines. However, the line aspect ratio plays a dominant role in affecting the thermal stresses. When the lower-level lines have a higher aspect ratio than the upper-level lines, the stresses in lower-level lines are significantly higher. This is likely to be the case in actual devices where typically many lower-level lines are also narrower (see Fig. 1.4). Analyses considering the simpler two-level arrangement have also been reported [84, 100].

4.4.2 Copper Lines and Via

As schematically shown in Fig. 4.1, different levels of metal lines are connected through the via. The via and its vicinity are frequently susceptible to voiding failure. There have been numerical analyses on Al and Cu interconnects focusing on the via region [100–112]. In the following we consider a case study, with attention directed to via deformation and the effect of dielectric materials in the Cu interconnect system. Figure 4.16 shows the model geometry. The direction of the elastic-plastic Cu lines is parallel to the x-axis. The computational domain represents one half of the unit structure, with the xz-plane exposed in Fig. 4.16 being the mirror symmetry plane showing the middle cut of the metal line/via structure. The model is bound by a top and a bottom layer silicon oxide (SiO_2). The thin barrier layers directly on top of both levels of Cu lines are silicon nitride (SiN_x) and the thin side/bottom barrier layers bonding Cu are tantalum nitride (TaN). The Cu via is assumed to be a rectangular block with the same width (in the y-direction) as the Cu lines. Three different regions of the dielectric, having the same vertical positions as the lower line, via, and upper line, are defined. In the present example all three dielectric regions are filled with the same material, either SiO_2, or carbon-doped glass-based low-k material SiCOH, or BCB low-k material. A cooling process from

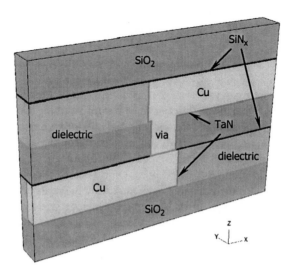

Fig. 4.16 Three-dimensional model geometry used for simulating dual level Cu lines connected by the via. The entire model has spans of 3.0, 0.28 and 2.0 μm along the x, y and z directions, respectively. The *top* and *bottom* SiO$_2$ layers are 0.38-μm thick. The thicknesses of all the nitride barrier layers are 0.02 μm. The height of both levels of Cu lines is 0.4 μm, and the width is 0.24 μm (only 0.12 μm included in the computational domain due to symmetry). The Cu via, located at the center in the x direction, is assumed to be a rectangular block with the same width (in the y direction) as the Cu lines. The via dimensions in the x and z directions, not including the barrier layer, are 0.3 and 0.38 μm, respectively. Three cases of the dielectric material are considered: SiO$_2$, SiCOH and BCB polymer [91]

350 to 20°C is simulated. In the modeling the bottom xy-plane is clamped on a rigid substrate. All lateral boundaries in Fig. 4.16 are mirror symmetry planes: they are constrained to remain vertical while the nodal points in the respective planes are allowed to undergo in-plane movements during deformation. The top surface is not constrained. All material properties used in the modeling are listed in Table 4.1. In the numerical implementation, the CTE values used for all materials are those relative to the CTE of the silicon substrate (3.0×10^{-6} K^{-1}), which is not explicitly included in the model.

Figure 4.17a, b and c show the contour plots of hydrostatic stress in Cu after cooling, for the cases of the traditional SiO$_2$ dielectric, SiCOH low-k dielectric, and BCB low-k dielectric, respectively. Materials other than Cu in the model are not displayed in these figures. It can be seen that the tensile stress is generally much lower for the two cases of low-k dielectric than for the oxide dielectric. The reduction is particularly significant in the via and its vicinity. In fact, the via in Fig. 4.17c is under compression due to the fact the stress component σ_{zz} is highly compressive (caused mainly by the high CTE of the polymer low-k surrounding the via). If one uses the hydrostatic tensile stress as a metric for predicting damage propensity, then the results in Fig. 4.17 will not be consistent with the typical failure pattern observed in actual interconnects where the via

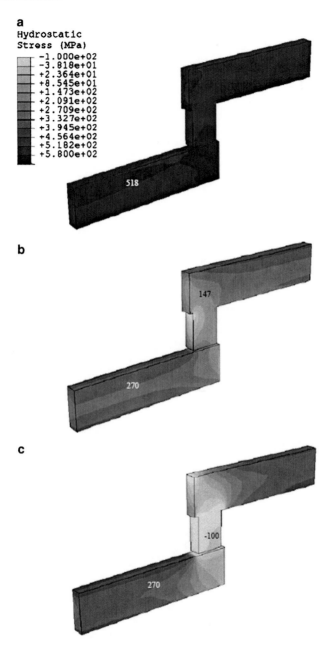

Fig. 4.17 Contours of hydrostatic stress in Cu after cooling. The dielectric materials used in the model are (**a**) SiO$_2$, (**b**) SiCOH, and (**c**) BCB polymer

region is easily subject to voiding and interfacial debonding. This is because an important piece of information, namely the plastic deformation in Cu, has not been considered.

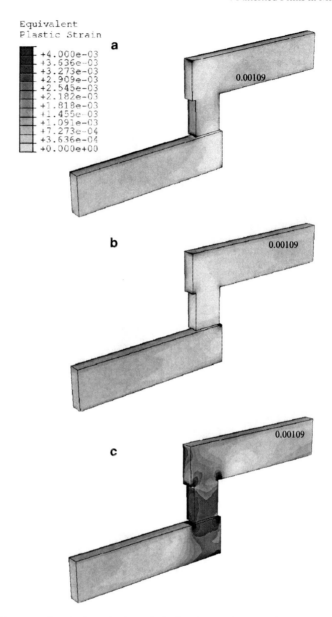

Fig. 4.18 Contours of equivalent plastic strain in Cu after cooling. The dielectric materials used in the model are (**a**) SiO$_2$, (**b**) SiCOH, and (**c**) BCB polymer

Figure 4.18a, b and c show the contour plots of equivalent plastic strain in Cu after cooling, for the cases of the traditional SiO$_2$ dielectric, SiCOH low-k dielectric, and BCB low-k dielectric, respectively. It is seen that plastic deformation is more concentrated near the interfaces, especially in the corner regions of the via.

The plastic strain is particularly high in the case of the BCB low-*k* dielectric, which is a natural consequence of the high deviatoric stress field. This brings about the possibility of severe shearing (permanent distortion) of the via. Such type of failure mode has been reported [113–115], where discrete shearing of the upper part of the via relative to the lower part creates highly discernible shear steps at the interface or shear cracks through the via. The mechanically weak low-*k* material is unable to prevent this from occurring. The use of SiCOH low-*k* dielectric (having the CTE and modulus values between those of silicon oxide and polymer), however, tends to alleviate the problem. The plastic strains are comparable to those with the oxide dielectric (Fig. 4.18b), and the hydrostatic tension is much reduced compared with the case of oxide dielectric (Fig. 4.17b). The plastic strain contours shown in Fig. 4.18 may also explain other experimentally observed voiding phenomena including void formation at the via bottom [22, 111], near the trench shoulder area [116], and along the top interface of the lower-level Cu [22, 106]. Within the simple continuum framework, the analysis of constrained deformation is seen to yield meaningful and relevant information regarding the structural integrity of Cu interconnects.

4.4.3 Stresses in Barrier Layers and Dielectrics

Due to the generally weak mechanical strength of low-*k* dielectric materials, there have been concerns regarding the stresses "passed on" to the thin barrier layers. Because of their brittle nature and thin dimensions, one appropriate stress parameters relating to potential failure is the maximum principal stress (if it is tensile). Figure 4.19a, b and c show the contour plots of maximum principal stress in the SiN_x and TaN layers for the cases of SiO_2, SiCOH and BCB dielectrics, respectively. For clarity all top and bottom SiO_2, dielectric and Cu materials are removed from the presentation. Note that the contours shown in Fig. 4.19 do not provide information on the principal directions of the stress tensor, but the majority portion of the barrier layers is seen to subject to tensile stresses. The local tensile stress can reach a high magnitude (well above 800 MPa). In Fig. 4.19a, the TaN layer at the via bottom and the lower SiN_x layer adjacent to the via (partially obstructed from the view) are seen to be under relatively high stresses. In the case of SiCOH dielectric (Fig. 4.19b), the stresses are generally smaller except along a narrow strip in the lower SiN_x layer. The use of polymer-based dielectric results in high stresses in the narrow region of lower SiN_x as well as in the upper SiN_x in regions away from the upper metal line (Fig. 4.19c). The very high tensile stresses in the upper SiN_x is primarily due to the enhanced contraction in the vertical (*z*) direction of the BCB polymer upon cooling, causing downward deflection of the SiN_x beam near the left boundary. This may lead to easy initiation of brittle fracture. (The right half of upper SiN_x can be thought of as clamped because it is supported by a stiffer structure of the metal line.)

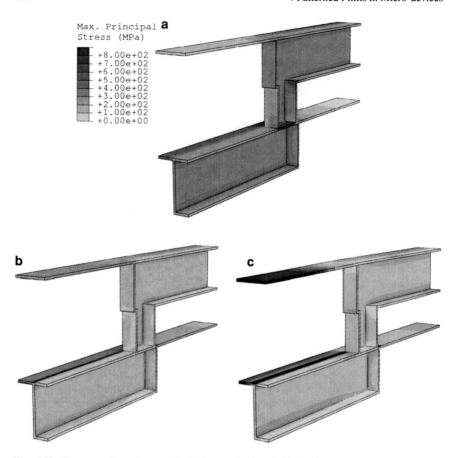

Fig. 4.19 Contours of maximum principal stress in the nitride barrier layers after cooling. The dielectric materials used in the model are (**a**) SiO_2, (**b**) SiCOH, and (**c**) BCB polymer

Since low-k dielectrics have very different mechanical properties from typical ceramic materials, it is of interest to quantitatively examine the stresses in the low-k dielectrics themselves. Figure 4.20a and b show the contour plots of hydrostatic stress in the SiCOH and BCB low-k dielectric materials, respectively. In the figures all other materials are removed from the presentation. It can be seen that, in both models, the highest tensile stresses appear in regions adjacent to the side walls of the metal lines at both levels. In a global sense, these are the regions where the dielectric is tightly constrained by the series of parallel metal lines. The hydrostatic tension in the BCB polymer can be well over 100 MPa, while in the much stiffer SiCOH the stress is actually lower. Note that a tensile stress level of over 100 MPa may be tolerated by typical metallic and ceramic materials, but great uncertainty exists for polymer-based systems. The results presented in Figs. 4.19 and 4.20 suggest that, for thermal stress concerns, using SiCOH is a "safer" approach in the copper interconnect/low-k dielectric integration.

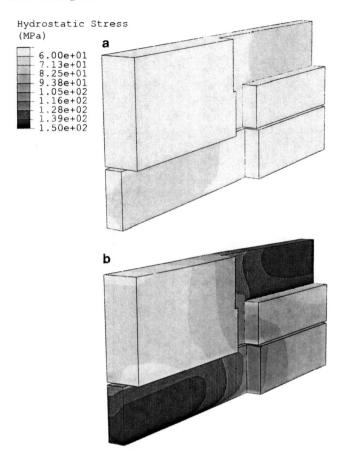

Hydrostatic Stress
(MPa)

- 6.00e+01
- 7.13e+01
- 8.25e+01
- 9.38e+01
- 1.05e+02
- 1.16e+02
- 1.28e+02
- 1.39e+02
- 1.50e+02

Fig. 4.20 Contours of hydrostatic stress in the (**a**) SiCOH low-*k* dielectric and (**b**) BCB low-*k* dielectric, after cooling. In the case of SiO_2 dielectric, the hydrostatic stress is largely compressive with magnitudes smaller than 100 MPa (not shown here)

4.5   Lines with Pre-existing Flaws

Defects caused by processing are commonly encountered in metal interconnects and other thin-film devices. For instance, there have been strong indications that voids in Al lines tend to form at the side-wall interface between the metal and dielectric, and the pre-existing interface flaws due to contamination such as organic etch residue serve as sites for void nucleation. Theoretical analyses within the context of classical nucleation have illustrated this possibility [6, 117, 118]. Experimental studies also showed that additional cleaning processes can dramatically suppress voiding in Al interconnects [119, 120].

4.5.1 Effect of Local Debond

When locally debonded areas exist, thermal stresses generated in the metal lines upon cooling from the processing temperature differ from those with perfectly bonded interfaces. Numerical modeling of debond-mediated stresses in Al intercon-nects has been undertaken [121, 122]. We now present 3D simulation results, the model of which is shown in Fig. 4.21a. The Al line is embedded within the SiO_2 dielectric, with the x-direction representing the longitudinal direction. The bottom surface ($z=0$) is assumed to be clamped on a substrate, with displacements in y and z prohibited but allowed in the x-direction. Mirror symmetry across the front ($x=0$), back ($x=l/2$) and side surfaces ($y=p/2$ and $y=-p/2$) is assumed for simulating a periodic structure. Following the case in Sect. 4.2.1, the thicknesses of the SiO_2 layer directly above and below Al are taken to be the same, and the magnitude of the total SiO_2 height, t, is assumed to be twice that of the line height, h. The aspect ratio (h/w) and the pitch-to-width ratio (p/w) of the Al line are taken to be 1 and 2, respectively. The local debonded area is set to be a pre-existing free surface. The debond was taken to span the entire line height on one side-wall interface, with the debond center located at $x=0$ and the debond length denoted by l_d. The magnitude of l is taken to be at least four times of l_d in the calculations. The thermal history considered is from 400°C to room temperature. The material properties are listed in Table 4.1.

Figure 4.21b shows the predicted normal and hydrostatic stresses, averaged over the cross-section area of the Al line, as a function of position along the line direc-tion, with a debond length l_d twice of the line height h. The distance in x is normal-ized by h, so the debond center and debond edge are represented by the coordinates 0 and 1, respectively, of the abscissa. Also indicated in the figure are the calculated stress levels without any interfacial debond, which are independent of the position along the line. The presence of a debond segment induces a highly non-uniform stress field. The trend of the stress variation along the line, however, can be cap-tured clearly by the area-averaged values. It can be seen that the lowest tensile stresses occur near the debond center. The stresses increase rapidly along the line direction near the debond edge. At about a distance of one line height (h) away from the edge of debond along the line, the magnitudes of all stress components essen-tially resume those in the case without any pre-existing debond.

For the three normal stress components, σ_{yy} shows the greatest extent of reduc-tion at the debond center (from about 580 to 110 MPa). This is because the traction-free boundary condition on the crack (debond) surface needs to be satisfied. The magnitude of σ_{yy} is greater than that of σ_{zz} outside the debond edge, but the opposite is true within the debonded segment. The stress σ_{xx} is the highest component within the debonded segment and along most of the intact parts of the line. However, in regions immediately outside the debond edge, σ_{yy} is greater than σ_{xx}. The physical implication of this observation will be discussed below.

The same qualitative features are also found for different debond lengths. To easily compare the stress gradients along the line direction, the stress–distance curves are now plotted in Fig. 4.22 with the debond edge taken as the origin.

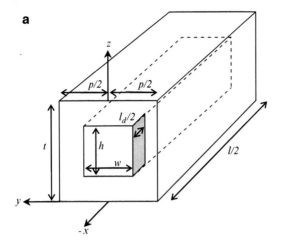

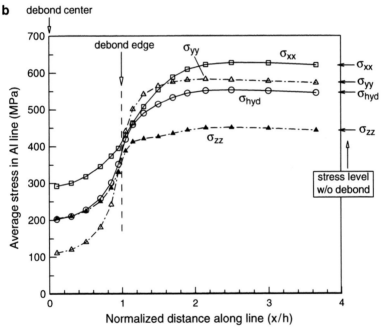

Fig. 4.21 (a) Schematic of the debonded interconnect used in the numerical modeling. The *shaded area* represents the debond. (b) Variation of the stresses in Al (averaged over the cross-section area of the Al line) along the line direction, with the presence of a pre-existing debond segment of length $2h$ [121]

Figure 4.22a and b show the profiles of hydrostatic stress and equivalent plastic strain, respectively, for the debond lengths of $0.5h$, h and $2h$. Also included in the figures are the values in the case without any debond, which are constant along the line. It is seen from Fig. 4.22a that, while the stress magnitudes within the

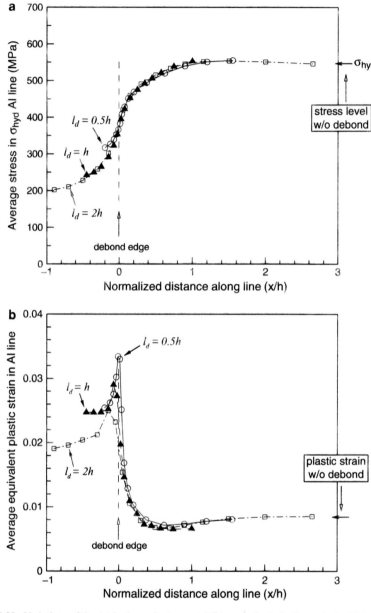

Fig. 4.22 Variations of the (**a**) hydrostatic stress and (**b**) equivalent plastic strain, in Al (averaged over the cross-section area of the Al line) along the line direction, with the presence of a pre-existing debond segment of lengths $0.5h$, h, and $2h$. The origin of x is set to be at the debond edge [121]

debonded segment differ, their variations with the line distance outside the debond are essentially the same for the three debond lengths considered. The stress magnitudes right at the debond edge are approximately 65–70% of those far away from the debond. The stress gradient essentially vanishes at locations beyond a distance of h from the debond edge. The plastic strain distribution in Fig. 4.22b shows significant higher values near the debond segment than in the line interior (away from the debond). The plastic strain magnitudes essentially resume that of the case without any debond at less than a distance of h away from the debond edge. The maximum plastic strain occurs very close to the debond edge within the debonded segment, and its value decreases as the debond length increases. The enhancement of plasticity near the debond is clearly associated with the reduction of the triaxial constraint, due to the relatively free movement of the debond surface upon thermal contraction.

It should be noted that, as mentioned above, the stress and strain distributions presented in Figs. 4.21 and 4.22 are cross-section averaged values. On any cross section, non-uniformity of field quantities exists, especially in areas near the debond. Nevertheless, salient features are provided by the variation of averaged stress and strain values along the line direction as shown. The calculated stress profile can be applied to void growth models [123, 124] where quantitative information on the distance of stress relaxation along the line is required. The in-line stress distribution also serves as the initial condition when applying the one-dimensional numerical models [125–129] to simulate the electromigration stress buildup, if the pre-existing debond is to be accounted for [130, 131].

4.5.2   Deformation Induced Void Opening

The predicted stress profile along a locally debonded line presented above can be used to facilitate a micromechanical model of void formation *during cooling*, which is based on the crystallographic slip in the metal [121, 132, 133]. (The atomic diffusion contribution is negligible within the time scale encountered in normal thermal cooling operation [134–136].) Experiments have shown that voids in straight metal lines nucleate and grow from the side interface without significant morphological changes along the vertical (z) direction. Therefore, as a first approximation, it suffices to consider only the stress components along the xy-plane for the discussion of slip geometry and voiding behavior. Moreover, for simplicity it is assumed that the slip systems of the metal are along the 45° directions, symmetrically oriented with respect to the line direction. Figure 4.21b illustrated that within the debonded segment, σ_{xx} is greater than σ_{yy}. Immediately outside the debond edge, there exists a line segment within which σ_{yy} is greater than σ_{xx}. The length of this segment is approximately 0.5–0.6h. Figure 4.23a delineates the stress state in the vicinity of the debond using the conventional representation in elementary mechanics. Note that this corresponds to the "top view" of the interconnect structure, with the xy-plane assumed to be the wafer plane. Thus, the horizontal metal/dielectric interfaces represent the vertical side walls of the metal line. The shear stress state along the 45° directions, in accord with the normal stress state in Fig. 4.23a, can now be deduced and is shown schematically in Fig. 4.23b. The same feature holds true for other debond lengths considered in the model [121]. Figure 4.23c shows the crystallographic model.

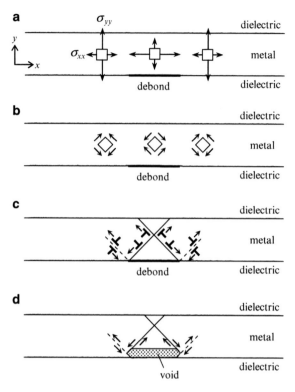

Fig. 4.23 Modes of the (**a**) normal and (**b**) shear components of stress in the metal line near the pre-existing debond upon cooling. The slip of edge dislocations in response to the shear mode in (**b**) is schematically shown in (**c**). Slip systems denoted by the *solid* and *dashed lines* represent slip operations under different shear modes. The dislocation slip activities at the debond site give rise to the displacement of matter and, therefore, void opening in (**d**) [121]

Plastic yielding occurs by slip of edge dislocations along the slip systems in this 2D model. The "extra half plane" associated with the dislocation is specified by the conventional dislocation symbols in typical materials science textbooks. The dislocation motion specified in Fig. 4.23c is *fully compatible* with the shear stress state shown in Fig. 4.23b. Although only the edge dislocations are included, the same kind of slip behavior can also be accomplished by the slip of screw and mixed dislocations. The consequences of the slip activities in Fig. 4.23c are depicted in Fig. 4.23d, where the block movement of matter in response to the shearing is shown. Void opening from the debond site is apparent, and the local elastic thermal stresses can be partially relieved.

The above discussion demonstrated that, aside from its practical applicability, continuum-based thermo-mechanical modeling of interconnects can also play a part in gaining fundamental insight into the physics of reliability problems. In addition, the modeling result may be used to define appropriate boundary conditions for atomistic simulations, for examining the interrelation between crystal defect mechanisms and void formation at the atomic scale.

4.6 Voiding Damage and Stress Relaxation

The formation of voids is a common failure mechanism in metal interconnects in microelectronic devices. Since the encapsulated metal line is under high tensile stresses after cooling, voiding is in fact a natural way of relaxing the stress. The misfit volumetric strain is commonly expressed as the magnitude of $3(\alpha_{metal} - \alpha_{surr})\Delta T$, where α_{metal} is the CTE of the metal and α_{surr} stands for the effective CTE of materials surrounding the metal. It is often conceived that the stress in the metal line will relax *completely* if the total void volume fraction attains the same value of $3(\alpha_{metal} - \alpha_{surr})\Delta T$, which may be termed "saturation void fraction" [137, 138]. If $(\alpha_{metal} - \alpha_{surr})$ is taken to be 2.2×10^{-5} K^{-1} for Al interconnects and the magnitude of ΔT is taken as 380 K, the saturation void fraction becomes 0.025. We can use a simple finite element analysis to "test" this perception of saturation fraction, as illustrated in Sect. 4.6.1 below.

4.6.1 Voiding Induced Stress Relaxation

Figure 4.24a shows the 2D model of an Al line embedded within the SiO$_2$ dielectric used for the modeling. Conceptually the *xy*-plane is parallel to the wafer plane, and the Al/SiO$_2$ interfaces are the vertical side walls of the Al interconnect. The length *l* is taken to be equal to 8*w*, where *w* is the line width. The sliding boundary condition is imposed on the four external boundaries, where no displacement perpendicular to the boundary line is allowed. As a consequence, mirror symmetry exists across the side boundaries. The generalized plane strain condition is employed, with a uniform strain along the z-direction superimposed on the plane deformation state. This partially relieves the plane strain constraint which may be a significant source of error. With the present simplified model, reasonably accurate stress and strain fields can still be obtained [130]. A thermal cooling process from 400 to 20°C is first imposed on the structure free of any void. Voiding is then simulated by removing targeted elements under the static equilibrium condition. Note that the choice of the void size and shape is arbitrary; in this example we create a void with trapezoidal shape and the void volume fraction of 0.031. The purpose here is not to simulate failure, but to examine how the stress field will be altered as a consequence of voiding. The material properties used in the modeling are listed in Table 4.1. In the numerical implementation, the CTE values used for all materials are those relative to the CTE of the silicon substrate (not specifically included in the model).

Figure 4.24b and c show the contour plots of hydrostatic stress after cooling and after subsequent void creation, respectively. In the absence of any void, the stress generated in the metal line has a uniform value of 706 MPa. In Fig. 4.24c the local relief of stress near the void is evident. It should be noted that the stress does not vanish at the void site, because the traction-free condition on the void surface does

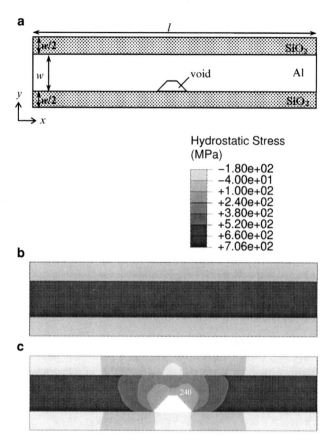

Fig. 4.24 (a) Schematic of the 2D interconnect model and the intended void of *trapezoidal shape*. Contour plots of hydrostatic stress (**b**) after cooling from 400 to 20°C and (**c**) after the creation of a void

not eliminate the "in-surface" normal components of the stress tensor. The stress increases with the distance away from the void site, which results in a stress gradient on both sides of the void. The main point of this analysis, however, is that, while the void volume fraction already exceeds the "saturation value," the stress has relaxed only partially in the vicinity of the void. In fact the *volume averaged* stress in the entire Al line in Fig. 4.24c is still at about 75% of the stress magnitude without the presence of a void. Even if the Al/SiO$_2$ interface is allowed to slip freely without normal separation, no significant further relaxation of the average stress will occur since the overall confinement of the metal line in all dimensions is still the same. The concept of saturation void fraction is only valid for materials capable of accommodating shape changes in an *unconstrained* manner. This is not the case in encapsulated metal interconnects in which thermal stresses cannot fully relax through voiding *alone*.

4.6.2   Time-Dependent Deformation and Void Growth

A possible counterargument to the numerical illustration in Sect. 4.6.1 is that time-dependent material response, capable of relaxing stress over time, should be involved in assessing the correlation between stress relaxation and voiding. To explore this, the 2D interconnect geometry identical to that in Fig. 4.24 is used, with the time-dependent creep properties of Al accounted for. The creep part of the material model follows that of "diffusional creep" in polycrystalline pure Al [139] (see also (3.12) for thin continuous films under the equi-biaxial state). Although its applicability to fine interconnect lines is in doubt, the purpose here is to numerically experimenting qualitative features for demonstration, not to make quantitative predictions. The simulation assumes a thermal history of fast cooling from 400 to 20°C and then heating back to 250°C (presumably the testing temperature), which is followed by the creation of a "seed void" through element removal while maintaining the static equilibrium condition. The small artificial void is in the form of interfacial slit, which corresponds to a void volume fraction of 0.0003. Note that this volume fraction is very small compared to the supposed saturation void fraction at 250°C (about 0.01). It serves as a starting mechanism for simulating subsequent void growth. The structure is then left undergoing stress relaxation at 250°C.

Figure 4.25a to d show the contour plots of hydrostatic stress in the structure after 0, 100, 200 and 500 s, respectively, during the relaxation. It can be seen that the stress value near the void is lower compared to regions away from the void. With increasing time the low-stress region expands, and the magnitude of stress continues to decrease throughout the entire metal line. Accompanied by the stress relaxation is the continued increase in void size. Void growth in the model is *self-occurring*, induced by the relaxing metal without any failure criterion imposed. At 500 s the hydrostatic stress has reduced to below 100 MPa in the entire Al line. By this time the void size is calculated to be about 3.5 times that of the supposed saturation value at 250°C. Evidently the conceived saturation limit still does not exist.

Within the context of correlating interconnect stress with voiding damage, there is another issue worthy of addressing here. Some researchers have utilized the spatial gradient of hydrostatic stress obtained from 3D finite element modeling to explain voiding propensity in interconnects [108, 109, 112, 140, 141]. It should be noted that the hydrostatic stress gradient is related to the vacancy flux, which can be treated as a driving force for the *growth* of an existing void. But in these studies an *intact* metal line structure was used. If a void has been nucleated (or pre-exiting upon fabrication), then the stress field would be disturbed so the modeled stress gradient would no longer be valid. As a consequence, one should exercise care when interpreting the modeling result. The gradient of hydrostatic stress can be used for rationalizing the void growth tendency only when an existing void is incorporated in the model such as in Refs. [111, 142]. Furthermore, an appropriate elastic-plastic model for the metal has to be used to ensure the accuracy of the simulated stress field.

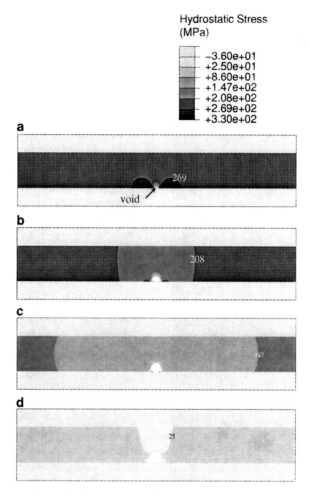

Fig. 4.25 Contour plots of hydrostatic stress after (**a**) 0 s, (**b**) 100 s, (**c**) 200 s and (**d**) 500 s, during isothermal stress relaxation at 250°C. With a small seed void introduced in the model, void growth is self-occurring without any failure criterion imposed. The creep model used is $\varepsilon = 2.064 \times 10^{-5}\sigma$ (strain rate unit: s^{-1}; stress unit: MPa), which is based on the diffusional creep mechanism for polycrystalline aluminum with an average grain size of 1 μm [139]

4.7 Projects

1. Consider the modeling of parallel Al lines on a Si substrate as in Sect. 4.1.2. Apply the same approach to lines with different aspect ratios and pitch-to-width ratios over a wide range. Examine the local stress field σ_{xx}, σ_{yy}, σ_{zz} and σ_{yz}. Calculate the average stresses σ_{xx} and σ_{yy} in the lines and the overall curvatures k_x and k_y. Are you able to correlate σ_{xx} and σ_{yy} with k_x and k_y using fundamental mechanics relations? Discuss any trend that you can identify for both the cases of etching and thermal loading.

2. With regard to the structure of parallel lines on a Si substrate considered in Sects. 4.1.1 and 4.1.2, conduct similar numerical modeling but with anisotropic material properties. Familiarize yourself with the features of material anisotropy available in your finite element program. Carry out a parametric analysis with anisotropic yield behavior of Al and study how the stresses and curvature will be affected. What will be the effects if the anisotropic elastic properties of Si and Al are used?

3. Consider the structure of single-level, passivated, parallel Al lines embedded within the SiO_2 dielectric as in Sect. 4.2.1 (Fig. 4.8a). Conduct a series of analysis using various combinations of aspect ratio and pitch-to-width ratios of the lines, as well as different SiO_2 geometries including its thickness above and below the metal. Calculate the averaged stresses σ_{xx} and σ_{yy} in Al under these different conditions, and examine the local fields of σ_{xx}, σ_{yy}, σ_{zz}, σ_H and equivalent plastic strain using contour plots. Correlate your observation with the geometric and mechanical features of the models. If a thick Si substrate is included in the models, how different will your results be?

4. Consider the Cu interconnect models employed in Sect. 4.2.2 (Figs. 4.9 and 4.11). A trend in the microelectronics industry is to use thinner and thinner barrier layers between copper and the surrounding dielectrics. Carry out a series of finite element analysis with decreasing thicknesses of SiN_x and TaN. Examine the changes in stress and deformation fields inside the Cu line, as well as in the barrier layers themselves and the dielectric materials. Will there be any different trends between the cases of all-SiO_2 dielectric and polymer-based dielectric?

5. The numerical case studies in Sects. 4.1 and 4.2 all used the generalized plane strain formulation to simulate the long-line configuration. Conduct a series of modeling using both the plane strain and generalized plane strain formulations and compare the differences. Discuss how different the results will be for unpassivated and passivated lines, for Al and Cu lines, for different cross-section geometries, and for different dielectric materials in the case of Cu interconnects?

6. With regard to the analyses of L and T shaped lines in Sect. 4.3, perform a systematic 3D modeling with L and T shaped Cu interconnects. Pay attention to the effects of barrier layers and the different dielectric materials. Will the overall trends be similar to those found in Al lines? What about the stresses in the barrier layers themselves? How will the results be altered in the case of densely arrayed Cu lines (similar to those shown in Fig. 4.12).

7. With regard to the 2D copper interconnect model used in Sect. 4.2.2 (Figs. 4.9 and 4.11), extend it to include multiple levels of lines. Use several combinations of line aspect ratios and pitch-to-width ratios. Also, try some cases with smaller sized lines at lower levels and larger sized lines at upper levels (see Fig. 1.4 but ignore vias). Will the stress field in the Cu lines be affected by the presence of lines at other levels? What about stresses in the barrier layers? How will the different dielectric materials affect the results?

8. Construct a 3D two-level Cu interconnect model similar to that in Fig. 4.16. Apply mechanical loading (tension, compression, and/or shear in different directions) instead of thermal loading. Study the evolutions of stress and strain

fields in copper, barrier layers and dielectrics. Compare the salient findings with those from the thermal loading. What if the structure is under the influence of combined mechanical and thermal loading? (Note in actual devices the packaging structure can impose significant mechanical stresses on the silicon chip.)

9. Construct a 3D two-level Cu interconnect model similar to that in Fig. 4.16. However, use a cylindrically shaped via instead of a rectangular block. How will the stress and strain fields be influenced by the via shape? Examine both the cases of mechanical loading (Project 8) and thermal loading.

10. Construct a 3D two-level Cu interconnect model similar to that in Fig. 4.16. Include one or more debonded segments (or slit-like voids at interfaces) in the model at different locations. Examine how the thermal stresses in Cu, barrier layers, and dielectrics can be affected by the debond patches.

11. Construct a 3D two-level Cu interconnect model similar to that in Fig. 4.16. Choose a creep model for Cu and apply it to your finite element analysis. How will the stress and deformation fields evolve as a result of thermal cooling? Next, with reference to the modeling illustration in Sect. 4.6.2, include a small "seed void" at certain locations in the Cu line or via in you model (e.g., corner or edge regions). Examine the stress relaxation as well as void growth behavior.

12. To achieve better insulation between the densely packed copper lines in advanced devices, the semiconductor industry has developed the "air gap" technology, taking advantage of the fact that vacuum is the best insulator available. The so-called air gaps are in fact voids (with low-pressure residual gas) in the bulk dielectric material. These empty spaces are adjacent to the side walls of the metal line, forming a tunnel-like structure parallel to the metal. Conduct literature and Internet searches, and select a representative configuration of the air gaps for inclusion in a finite element model. The model may be a modification of that in Fig. 4.9a. Simulate cooling and then thermal cycling, and examine the evolution of stress and deformation fields in the copper, barrier layers and dielectric materials. Compare the results with those without the air gaps. You may also investigate the effect of air gaps in a multilevel interconnect structure. Identify potential issues which may influence the thermo-mechanical reliability.

13. To further enhance the feature density and functionality of microelectronic devices, the 3D integration of interconnects and packaging has been under development. Survey the literature (e.g., [143–145]) and identify the possible thermo-mechanical issues associated with the 3D integration. Design a reasonable model and carry out a comprehensive numerical study.

References

1. V. P. Atluri, R. V. Mahajan, P. R. Patel, D. Mallik, J. Tang, V. S. Wakharkar, G. M. Chrysler, C.-P. Chiu, G. N. Choksi and R. S. Viswanath (2003) "Critical aspects of high-performance microprocessor packaging," MRS Bulletin, vol. 28(1), pp. 21–34.
2. S. Wolf (1990) Silicon processing for the VLSI era, Vol. 2 – process integration, Lattice Press, Sunset Beach.

3. M. Ohring and J. R. Lloyd (2009) Reliability and failure of electronic materials and devices, 2nd ed., Academic Press, San Diego.

4. P. A. Flinn, A. S. Mack, P. R. Besser and T. N. Marieb (1993) "Stress-induced void formation in metal lines," MRS Bulletin, vol. 18(12), pp. 26–35.

5. H. Okabayashi (1993) "Stress-induced void formation in metallization for integrated circuits," Materials Science and Engineering R, vol. R11, pp. 191–241.

6. P. A. Flinn (1995) "Mechanical stress in VLSI interconnects: Origins, effects, measurement, and modeling," MRS Bulletin, vol. 20(11), pp. 70–73.

7. J. R. Lloyd and J. J. Clement (1995) "Electromigration in copper conductors," Thin Solid Films, vol. 262, pp. 135–141.

8. T. D. Sullivan (1996) "Stress-induced voiding in microelectronic metallization: Void growth models and refinements," Annual Review of Materials Science, vol. 26, pp. 333–364.

9. E. Arzt, O. Kraft, R. Spolenak and Y.-C. Joo (1996) "Physical metallurgy of electromigration: Failure mechanisms in miniaturized conductor lines," Zeitschrift für Metallkunde, vol. 87, pp. 934–942.

10. A. S. Oates (1996) "Electromigration failure of contacts and vias in sub-micron integrated circuit metallizations," Microelectronics Reliability, vol. 36, pp. 925–953.

11. J. R. Lloyd (1997) "Electromigration in thin film conductors," Semiconductor Science and Technology, vol. 12, pp. 1177–1185.

12. D. G. Pierce and P. G. Brusius (1997) "Electromigration: A review," Microelectronics Reliability, vol. 37, pp. 1053–1072.

13. D. W. Malone and R. E. Hummel (1997) "Electromigration in integrated circuits," Critical Review in Solid State and Materials Science, vol. 22, pp. 199–238.

14. I. A. Blech (1998) "Diffusional back flows during electromigration," Acta Materialia, vol. 46, pp. 3717–3723.

15. S. H. Kang and J. W. Morris, Jr. and A. S. Oates (1999) "Metallurgical techniques for more reliable integrated circuits," JOM, vol. 51(3), pp. 16–18.

16. J. R. Lloyd, J. Clemens and R. Snede (1999) "Copper metallization reliability," Microelectronics Reliability, vol. 39, pp. 1595–1602.

17. C. Ryu, K.-W. Kwon, A. L. S. Loke, H. Lee, T. Nogami, V. M. Dubin, R. A. Kavari, G. W. Ray and S. S. Wong (1999) "Microstructure and reliability of copper interconnects," IEEE Transactions on Electron Devices, vol. 46, pp. 1113–1120.

18. S. M. Merchant, S. H. Kang, M. Sanganeria, B. van Schravendijk and T. Mountsier (2001) "Copper interconnects for semiconductor devices," JOM vol. 53(6), pp. 43–48.

19. E. T. Ogawa, K. D. Lee, V. A. Blaschke and P. S. Ho (2002) "Electromigration reliability issues in dual-damascene Cu interconnections," IEEE Transactions on Reliability, vol. 51, pp. 403–419.

20. Y.-L. Shen (2003) "Thermomechanical modeling of metal interconnects in microelectronic devices," in Recent research development in materials science VI, Research Signpost, Trivandrum, pp. 125–155.

21. C. S. Hau-Ridge (2004) "An introduction to Cu electromigration," Microelectronics Reliability, vol. 44, pp. 195–205.

22. B. Li, T. D. Sullivan, T. C. Lee and D. Badami (2004) "Reliability challenges for copper interconnects," Microelectronics Reliability, vol. 44, pp. 365–380.

23. Zs. Tokei, Y.-L. Li and G. P. Beyer (2005) "Reliability challenges for copper low-k dielectrics and copper diffusion barriers," Microelectronics Reliability, vol. 45, pp. 1436–1442.

24. M. Brillouet (2006) "Challenges in advanced metallization schemes," Microelectronics Reliability, vol. 83, pp. 2036–2041.

25. C. M. Tan and A. Roy (2007) "Electromigration in ULSI interconnects," Materials Science and Engineering R, vol. 58, pp. 1–75.

26. W. D. van Driel (2007) "Facing the challenge of designing for Cu/low-k reliability," Microelectronics Reliability, vol. 47, pp. 1969–1974.

27. M. A. Moske, P. S. Ho, D. J. Mikalsen, J. J. Cuomo and R. Rosenberg (1993) "Measurement of thermal stress and stress relaxation in confined metal lines. 1. Stresses during thermal cycling," Journal of Applied Physics, vol. 74, pp. 1716–1724.

28. I. S. Yeo, P. S. Ho and S. G. H. Anderson (1995) "Characteristics of thermal stresses in Al(Cu) fine lines. 1. Unpassivated line structures," Journal of Applied Physics, vol. 78, pp. 945–952.

29. Y.-L. Shen, S. Suresh and I. A. Blech (1996) "Stresses, curvatures, and shape changes arising from patterned lines on silicon wafers," Journal of Applied Physics, vol. 80, pp. 1388–1398.

30. M. J. Kobrinsky, C. V. Thompson and M. E. Gross (2001) "Diffusional creep in damascene Cu lines," Journal of Applied Physics, vol. 89, pp. 91–98.

31. A. Witvrouw, J. Proost, Ph. Roussel, P. Cosemans and K. Maex (1999) "Stress relaxation in Al-Cu and Al-Si-Cu thin films," Journal of Materials Research, vol. 14, pp. 1246–1254.

32. U. Burges, H. Helneder, M. Schneegans, D. Beckers, M. Hallerbach, H. Schroeder and W. Schilling (1995) "Thermal stresses in passivated AlSiCu-lines from wafer curvature measurement," in Thin Films: Stresses and Mechanical Properties V, Materials Research Society Symposium Proceedings, vol. 356, pp. 423–429.

33. I. S. Yeo, S. G. H. Anderson, P. S. Ho and C. K. Hu (1995) "Characteristics of thermal stresses in Al(Cu) fine lines. 2. Passivated line structures," Journal of Applied Physics, vol. 78, pp. 953–961.

34. N. Singh, A. F. Bower, D. Gan, S. Yoon, P. S. Ho, J. Leu and S. Shankar (2004) "Numerical simulations and experimental measurements of stress relaxation by interface diffusion in a patterned copper interconnect structure," Journal of Applied Physics, vol. 97, 013539.

35. R. P. Vinci and J. J. Vlassak (1996) "Mechanical behavior of thin films," Annual Review of Materials Science, vol. 26, pp. 431–462.

36. P. A. Flinn and G. A. Waychunas (1988) "A new x-ray diffraction design for thin-film texture, strain, and phase characterization," Journal of Vacuum Science and Technology B, vol. 6, pp. 1749–1755.

37. P. A. Flinn and C. Chiang (1990) "X-ray diffraction determination of the effect of various passivations on stress in metal films and patterned lines," Journal of Applied Physics, vol. 67, pp. 2927–2931.

38. I. C. Noyan, J. Jordan-Sweet, E. G. Liniger and S. K. Kaldor (1998) "Characterization of substrate/thin-film interfaces with x-ray microdiffraction," Applied Physics Letters, vol. 72, pp. 3338–3340.

39. B. Greenebaum, A. I. Sauter, P. A. Flinn and W. D. Nix (1991) "Stress in metal lines under passivation: Comparison of experiment and finite-element calculations," Applied Physics Letters, vol. 58, pp. 1845–1847.

40. M. A. Marcus, W. F. Flood, R. A. Cirelli, R. C. Kistler, N. A. Ciampa, W. M. Mansfield, D. L. Barr, C. A. Volkert and K. G. Steiner (1994) "X-ray strain measurements in fine-line patterned Al-Cu films," in Materials Reliability in Microelectronics IV, Materials Research Society Symposium Proceedings, vol. 338, pp. 203–208.

41. P. R. Besser, S. Brennen and J. C. Bravman (1994) "An x-ray method for direct determination of the strain state and strain relaxation in micron-scale passivated metallization lines during thermal cycling," Journal of Materials Research, vol. 9, pp. 13–24.

42. W. M. Kuschke and E. Arzt (1994) "Investigation of the stresses in continuous thin films and patterned lines by x-ray diffraction," Applied Physics Letters, vol. 64, pp. 1097–1099.

43. L. Maniguet, M. Ignat, M. Dupeux, J. J. Bacmann and Ph. Normandon (1994) "X-ray diffraction determination of the effect of passivation on stress in patterned lines of tungsten," in Materials Reliability in Microelectronics IV, Materials Research Society Symposium Proceedings, vol. 338, pp. 241–246.

44. P. R. Besser, T. N. Marieb, J. Lee, P. A. Flinn and J. C. Bravman (1996) "Measurement and interpretation of strain relaxation in passivated Al-0.5%Cu lines," Journal of Materials Research, vol. 11, pp. 184–193.

45. I. De Wolf, M. Ignat, G. Pozza, M. Maniguet and H. E. Maes (1999) "Analysis of local mechanical stresses in and near tungsten lines on silicon substrate," Journal of Applied Physics, vol. 85, pp. 6477–6485.

46. N. Yamamoto and S. Sakata (1995) "Strain analysis in fine Al interconnections by x-ray diffraction spectrometry using micro x-ray beam," Japanese Journal of Applied Physics, Part 2, vol. 34, pp. L664–667.

47. P. C. Wang, G. S. Cargill III, I. C. Noyan and C.-K. Hu (1998) "Electromigration-induced stress in aluminum conductor lines measured by x-ray microdiffraction," Applied Physics Letters, vol. 72, pp. 1296–1298.

48. H. H. Solak, Y. Vladimirsky, F. Cerrina, B. Lai, W. Yun, Z. Cai, P. Ilinski, D. Legnini and W. Rodrigues (1999) "Measurement of strain in Al-Cu interconnect lines with x-ray microdiffraction," Journal of Applied Physics, vol. 86, pp. 884–890.

49. P. C. Wang, I. C. Noyan, S. K. Kaldor, J. Jordan-Sweet, E. G. Liniger and C. K. Hu (2001) "Real-time x-ray microbeam characterization of electromigration effects in Al(Cu) wires," Applied Physics Letters, vol. 78, pp. 2712–2714.

50. N. Tamura, R. S. Celestre, A. A. MacDowell, H. A. Padmore, R. Spolenak, B. C. Valek, N. M. Chang, A. Manceau and J. R. Patel (2002) "Submicron x-ray diffraction and its applications to problems in materials and environmental science," Review of Scientific Instruments, vol. 73, pp. 1369–1372.

51. S.-H. Rhee, Y. Du and P. S. Ho (2003) "Thermal stress characteristics of Cu/oxide and Cu/low-k submicron interconnect structures," Journal of Applied Physics, vol. 93, pp. 3926–3933.

52. S.-H. Rhee and P. S. Ho (2003) "Thermal stress characteristics of two-level Al(Cu) interconnect structure," Journal of Materials Research, vol. 18, pp. 848–854.

53. J.-M. Paik, H. Park, Y.-C. Joo and K.-C. Park (2005) "Effect of dielectric materials on stress-induced damage modes in damascene Cu lines," Journal of Applied Physics, vol. 97, 104513.

54. A. S. Budiman, W. D. Nix, N. Tamura, B. C. Valek, K. Gadre, J. Maiz, R. Spolenak and J. R. Patel (2006) "Crystal plasticity in Cu damascene interconnect lines undergoing electromigration as revealed by synchrotron x-ray microdiffraction," Applied Physics Letters, vol. 88, 233515.

55. Q. Ma, S. Chiras, D. R. Clarke and Z. Suo (1995) "High-resolution determination of the stress in individual interconnect lines and the variation due to electromigration," Journal of Applied Physics, vol. 78, pp. 1614–1622.

56. I. De Wolf (1996) "Micro-Raman spectroscopy to study local mechanical stress in silicon integrated circuits," Semiconductor Science and Technology, vol. 11, pp. 139–154.

57. S. A. Smee, M. Gaitan, D. B. Novotny, Y. Joshi and D. L. Blackburn (2000) "IC test structures for multilayer interconnect stress determination," IEEE Electron Device Letters, vol. 21, pp. 12–14.

58. B. J. Aleck (1949) "Thermal stresses in a rectangular plate clamped along an edge," Journal of Applied Mechanics, vol. 16, pp. 118–122.

59. I. A. Blech and A. A. Levi (1981) "Comments on Aleck's stress distribution in clamped plates," Journal of Applied Mechanics, vol. 48, pp. 442–445.

60. H. Niwa, H. Yagi, H. Tsuchikawa and M. Kato (1990) "Stress distribution in an aluminum interconnect of very large scale integration," Journal of Applied Physics, vol. 68, pp. 328–333.

61. M. A. Korhonen, R. D. Black and C.-Y. Li (1991) "Stress relaxation of passivated aluminum line metallizations on silicon substrates," Journal of Applied Physics, vol. 69, pp. 1748–1755.

62. A. Wikstrom, P. Gudmundson and S. Suresh (1999) "Thermoelastic analysis of periodic thin lines deposited on substrate," Journal of the Mechanics and Physics of Solids, vol. 47: 1113–1130.

63. A. Wikstrom, P. Gudmundson and S. Suresh (1999) "Analysis of average thermal stresses in passivated metal interconnects," Journal of Applied Physics, vol. 86, pp. 6088–6095.

64. A. Gouldstone, A. Wikstrom, P. Gudmundson and S. Suresh (1999) "Onset of plastic yielding in thin metal lines deposited on substrates," Scripta Materialia, vol. 41, pp. 297–304.

65. A. Wikstrom and P. Gudmundson (2000) "Stresses in passivated lines from curvature measurements," Acta Materialia, vol. 48, pp. 2429–2434.

66. T. S. Park and S. Suresh (2000) "Effects of line and passivation geometry on curvature evolution during processing and thermal cycling in copper interconnect lines," Acta Materialia, vol. 48, pp. 3169–3175.

67. P. Sharma, H. Ardebili and J. Loman (2001) "Note on the thermal stresses in passivated metal interconnects," Applied Physics Letters, vol. 79, pp. 1706–1708.
68. C. H. Hsueh (2002) "Modeling of thermal stresses in passivated interconnects," Journal of Applied Physics, vol. 92, pp. 144–153.
69. P. Gudmundson and A. Wikstrom (2002) "Stresses in thin films and interconnect lines," Microelectronic Engineering, vol. 60, pp. 17–29.
70. T.-S. Park, M. Dao, S. Suresh, A. J. Rosakis, D. Pantuso and S. Shankar (2008) "Some practical issues of curvature and thermal stress in realistic multilevel metal interconnect structures," Journal of Electronic Materials, vol. 37, pp. 777–791.
71. S. Timoshenko (1976) Strength of materials, 3rd ed., Krieger, Huntington, New York.
72. A. Gouldstone, Y.-L. Shen, S. Suresh and C. V. Thompson (1998) "Evolution of stresses in passivated and unpassivated metal interconnects," Journal of Materials Research, vol. 13, pp. 1956–1966.
73. J. C. Lambropoulos and S. M. Wan (1989) "Stress concentration along interfaces of elastic-plastic thin films," Materials Science and Engineering A, vol. 107, pp. 169–175.
74. A. I. Sauter and W. D. Nix (1992) "Thermal stresses in aluminum lines bonded to substrates," IEEE Transactions on Components, Hybrids and Manufacturing Technology, vol. 15, pp. 594–600.
75. Y. Zhang and M. L. Dunn (2009) "Patterned bilayer plate microstructures subjected to thermal loading: Deformation and stresses," International Journal of Solids and Structures, vol. 46, pp. 125–134.
76. J.-H. Zhao, W.-J. Qi and P. S. Ho (2002) "Thermomechanical property of diffusion barrier layer and its effect on the stress characteristics of copper submicron interconnect structures," Microelectronics Reliability, vol. 42, pp. 27–34.
77. R. E. Jones and M. L. Basehore (1987) "Stress analysis of encapsulated fine-line aluminum interconnect," Applied Physics Letters, vol. 50, pp. 725–727.
78. A. Saerens, P. Van Houtte and S. R. Kalidindi (2001) "Finite element modeling of microscale thermal residual stresses in Al interconnects," Journal of Materials Research, vol. 16, pp. 1112–1122.
79. Y.-L. Shen (1997) "Modeling of thermal stresses in metal interconnects: effects of line aspect ratio," Journal of Applied Physics, vol. 82, pp. 1578–1581.
80. G. L. Povirk, R. Mohan and S. B. Brown (1995) "Crystal plasticity simulations of thermal stresses in thin-film aluminum interconnects," Journal of Applied Physics, vol. 77, pp. 598–606.
81. D. Chidambarrao, K. P. Rodbell, M. D. Thouless and P. W. DeHaven (1994) "Line-width dependence of stress in passivated Al lines during thermal cycling," in Materials Reliability in Microelectronics IV. Materials Research Society Symposium Proceedings, vol. 338, pp. 261–268.
82. Y.-L. Shen and S. Suresh (1995) "Thermal cycling and stress relaxation response of Si-Al and Si-Al-SiO2 layered thin films," Acta Metallurgica et. Materialia, vol. 43, pp. 3915–3926.
83. Y.-L. Shen and U. Ramamurty (2003) "Constitutive response of passivated copper films to thermal cycling," Journal of Applied Physics, vol. 93, pp. 1806–1812.
84. Y.-L. Shen (1997) "Thermal stresses in multilevel interconnections: aluminum lines at different levels," Journal of Materials Research, vol. 12, pp. 2219–2222.
85. M. S. Kilijanski and Y.-L. Shen (2002) "Analysis of thermal stresses in metal interconnects with multilevel structures," Microelectronics Reliability, vol. 42, pp. 259–264.
86. C. K. Hu, R. Rosenberg and K. Y. Lee (1999) "Electromigration path in Cu thin-film lines," Applied Physics Letters, vol. 74, pp. 2945–2947.
87. C. S. Hau-Riege and C. V. Thompson (2001) "Electromigration in Cu interconnects with very different grain structures," Applied Physics Letters, vol. 78, pp. 3451–3453.
88. A. Roy, R. Kumar, C. M. Tan, T. K. S. Wong and C. H. Tung (2006) "Electromigration in damascence copper interconnects of line width down to 100 nm," Semiconductor Science and Technology, vol. 21, pp. 1369–1372.

89. J. R. Lloyd, M. W. Lane, E. G. Liniger, C. K. Hu, T. M. Shaw and R. Rosenberg (2005) "Electromigration and adhesion," IEEE Transactions on Device and Materials Reliability, vol. 5, pp. 113–118.

90. C. D. Hartfield, E. T. Ogawa, Y. J. Park, T. C. Chiu and H. L. Guo (2004) "Interface reliability assessment for copper/low-k products," IEEE Transactions on Device and Materials Reliability, vol. 4, pp. 129–141.

91. Y.-L. Shen (2008) "On the elastic assumption for copper lines in interconnect stress modeling," IEEE Transactions on Device and Materials Reliability, vol. 8, pp. 600–607.

92. P. R. Besser, Y.-C. Joo, D. Winter, M. V. Ngo and R. Ortega (1999) "Mechanical stresses in aluminum and copper interconnect lines for 0.18 μm logic technologies," in Materials Reliability in Microelectronics IX, Materials Research Society Symposium Proceedings, vol. 563, pp. 189–199.

93. R. Spolenak, N. Tamura, B. C. Valek, A. A. MacDowell, R. S. Celestre, H. A. Padmore, W. L. Brown, T. Marieb, B. W. Batterman and J. R. Patel (2002) "High resolution microdiffraction studies using synchrotron radiation," in Stress-Induced Phenomena in Metallization: Sixth International Workshop, pp. 217–228.

94. Website: www.dow.com/silk/lit/index.htm, The Dow Chemical Company. Website accessed September 18, 2009.

95. E. S. Ege and Y.-L. Shen (2003) "Thermomechanical response and stress analysis of copper interconnects," Journal of Electronic Materials, vol. 32, pp. 1000–1011.

96. Y.-L. Shen (1999) "Designing test interconnect structures for micro-scale stress measurement: An analytical guidance," Journal of Vacuum Science and Technology B, vol. 17, pp. 448–454.

97. A. S. Nandedkar, G. R. Srinivasan, J. J. Estabil and A. Domenicucci (1993) "Atomistic simulation of void nucleation in aluminum lines," Philosophical Magazine A, vol. 67, pp. 391–406.

98. H. A. Le, N. C. Tso, T. A. Rost and C.-U. Kim (1998) "Influence of W via on the mechanism of electromigration failure in Al-0.5Cu interconnects," Applied Physics Letters, vol. 72, pp. 2814–2816.

99. J. Lee and A. S. Mack (1998) "Finite element simulation of a stress history during the manufacturing process of thin film stacks in VLSI structures," IEEE Transactions on Semiconductor Manufacturing, vol. 11, pp. 458–464.

100. P. M. Igic and P. A. Mawby (2000) "Investigation of the thermal stress field in a multilevel aluminum metallization in VLSI systems," Microelectronics Reliability, vol. 40, pp. 443–450.

101. L. T. Shi and K. N. Tu (1994) "Finite-element modeling of stress distribution and migration in interconnecting studs of a three-dimensional multilevel device structure," Applied Physics Letters, vol. 65, pp. 1516–1518.

102. L. T. Shi and K. N. Tu (1995) "Finite-element stress analysis of failure mechanisms in a multilevel metallization structure," Journal of Applied Physics, vol. 77, pp. 3037–3041.

103. A. Mathewson, C. G. M. De Oca and S. Foley (2001) "Thermomechanical stress analysis of Cu/low-k dielectric interconnect schemes," Microelectronics Reliability, vol. 41, pp. 1637–1641.

104. V. Senez, T. Hoffmann, P. Le Duc and F. Murray (2003) "Mechanical analysis of interconnected structures using process simulation," Journal of Applied Physics, vol. 93, pp. 6039–6049.

105. J.-M. Paik, H. Park and Y.-C. Joo (2004) "Effect of low-k dielectric on stress and stress-induced damage in Cu interconnects," Microelectronic Engineering, vol. 71, pp. 348–357.

106. C. J. Zhai, H. W. Yao, A. P. Marathe, P. R. Besser and R. C. Blish II (2004) "Simulation and experiments of stress migration for Cu/low-k BEoL," IEEE Transactions on Device and Materials Reliability, vol. 4, pp. 523–529.

107. Y.-L. Shen (2005) "Analysis of thermal stresses in copper interconnects/low-k dielectric structures," Journal of Electronic Materials, vol. 34, pp. 497–505.

108. W. Shao, Z. H. Gan, S. G. Mhaisalkar, Z. Chen and H. Li (2006) "The effect of line width on stress-induced voiding in Cu dual damascene interconnects," Thin Solid Films, vol. 504, pp. 298–301.

109. Z. Gan, W. Shao, S. G. Mhaisalkar, Chen Z and H. Li (2006) "The influence of temperature and dielectric materials on stress induced voiding in Cu dual damascene interconnects," Thin Solid Films, vol. 504, pp. 161–165.

110. Y.-L. Shen (2006) "Thermo-mechanical stresses in copper interconnects – a modeling analysis," Microelectronic Engineering, vol. 83, pp. 446–459.

111. S. Orain, A. Fuchsmann, V. Fiori and X. Federspiel (2006) "Reliability issues in Cu/low-k structures regarding the initiation of stress-voiding or crack failure," Microelectronic Engineering, vol. 83, pp. 2402–2406.

112. S. Orain, J.-C. Barbe, X. Federspiel, P. Legallo and H. Jaouen (2007) "FEM-based method to determine mechanical stress evolution during process flow in microelectronics, application to stress-voiding," Microelectronics Reliability, vol. 47, pp. 295–301.

113. M. Fayolle, G. Passemard, M. Assous, D. Louis, A. Beverina, Y. Gobil, J. Cluzel and L. Arnaud (2002) "Integration of copper with an organic low-k dielectric in 0.12 μm node interconnect," Microelectronic Engineering, vol. 60, pp. 119–124.

114. T. M. Shaw, X.-H. Liu, C. Murray, M. Y. Wisniewski, G. Fiorenza, M. Lane, S. Chiras, R. R. Rosenberg, R. Filippi, J. Mcgrath, H. Rathore and V. Mcgahay (2003) "The mechanical behavior of low-k/copper interconnect structures," presentation at the 2003 Materials Research Society Fall Meeting, Boston, MA, U9.1.

115. R. G. Filippi, J. F. McGrath, T. M. Shaw, C. E. Murray, H. S. Rathore, P. S. McLaughlin et al. (2004) "Thermal cycle reliability of stacked via structures with copper metallization and an organic low-k dielectric," in Proceedings of the 42nd IEEE International Reliability Physics Symposium, pp. 61–67.

116. A. Sekiguchi, J. Koike and K. Maruyama (2002) "Formation of slit-like voids at trench corners of damascene Cu interconnects," Materials Transactions of JIM, vol. 43, pp. 1633–1637.

117. R. J. Gleixner, B. M. Clemens and W. D. Nix (1997) "Void nucleation in passivated interconnect lines: effects of site geometries, interfaces, and interface flaws," Journal of Materials Research, vol. 12, pp. 2081–2090.

118. P. A. Flinn, S. Lee, J. Doan, T. N. Marieb, J. C. Bravman and M. Madden (1998) "Void phenomena in passivated metal lines: recent observations and interpretation," in Stress Induced Phenomenon in Metallization, Fourth International Workshop, American Institute of Physics Conference Proceedings, vol. 418, pp. 250–261.

119. T. Wada, M. Sugimoto and T. Ajiki (1989) "Effects of surface treatment on electromigration in aluminum films," IEEE Transactions on Reliability, vol. 38, pp. 565–570.

120. H. Abe, S. Tanabe, Y. Kondo and M. Ikubo (1992) "The influence of adhesion between passivation and aluminum films on stress induced voiding," Extended Abstract in Japan Society of Applied Physics 39th Spring Meeting, p. 658.

121. Y.-L. Shen (1998) "Stresses, deformation, and void nucleation in locally debonded metal interconnects," Journal of Applied Physics, vol. 84, pp. 5525–5530.

122. Y.-L. Shen (1998) "Effects of pre-existing interfacial defects on the stress profile in aluminum interconnection lines," IEEE Transactions on Components, Packaging and Manufacturing Technologies, Part A, vol. 21, pp. 127–131.

123. F. G. Yost, D. E. Amos and A. D. Romig, Jr. (1989) "Stress-driven diffusive voiding of aluminum conductor lines," in Proceedings of the 27th IEEE International Reliability Physics Symposium, pp. 193–201.

124. W. D. Nix and A. I. Sauter (1992) "A study of stress-driven diffusive growth of voids in encapsulated interconnect lines," Journal of Materials Research, vol. 7, pp. 1133–1143.

125. M. A. Korhonen, P. Borgensen, K.-N. Tu and C.-Y. Li (1993) "Stress evolution due to electromigration in confined metal lines," Journal of Applied Physics, vol. 73, pp. 3790–3799.

126. B. D. Knowlton, J. J. Clement and C. V. Thompson (1997) "Simulation of the effects of grain structure and grain growth on electromigration and the reliability of interconnects," Journal of Applied Physics, vol. 81, pp. 6073–6080.

127. Y. J. Park and C. V. Thompson (1997) "The effects of the stress dependence of atomic diffusivity on stress evolution due to electromigration," Journal of Applied Physics, vol. 82, pp. 4277–4281.

128. Y. K. Liu, C. L. Cox and R. J. Diefendorf (1998) "Finite element analysis of the effects of geometry and microstructure on electromigration in confined metal lines," Journal of Applied Physics, vol. 83, pp. 3600–3608.

129. Q. F. Duan and Y.-L. Shen (2000) "On the prediction of electromigration voiding using stress-based modeling," Journal of Applied Physics, vol. 87, pp. 4039–4041.

130. Y.-L. Shen, Y. L. Guo and C. A. Minor (2000) "Voiding induced stress redistribution and its reliability implications in metal interconnects," Acta Materialia, vol. 48, pp. 1667–1678.

131. C. A. Minor, Y. L. Guo and Y.-L. Shen (1999) "On the propensity of electromigration void growth from preexisting stress-voids in metal interconnects," Scripta Materialia, vol. 41, pp. 347–352.

132. Y.-L. Shen (1997) "On the formation of voids in thin-film metal interconnects," Scripta Materialia, vol. 37, pp. 1805–1810.

133. Y.-L. Shen (1999) "Void nucleation in metal interconnects: combined effects of interface flaws and crystallographic slip," Journal of Materials Research, vol. 14, pp. 584–591.

134. C. A. Volkert, C. F. Alofs and J. R. Liefting (1994) "Deformation mechanisms of Al films on oxidized Si wafers," Journal of Materials Research, vol. 9, pp. 1147–1155.

135. M. D. Thouless, K. P. Rodbell and C. Cabral, Jr. (1996) "Effect of a surface layer on the stress relaxation of thin films," Journal of Vacuum Science and Technology A, vol. 14, pp. 2454–2461.

136. Y.-L. Shen and S. Suresh (1996) "Steady-state creep of thick and thin-film multilayers," in Polycrystalline Thin Films – Structure, texture, properties, and applications II, Materials Research Society Symposium Proceedings, vol. 403, pp. 133–138.

137. Z. Suo (1998) "Stable state of interconnect under temperature change and electric current." Acta Materialia, vol. 46, pp. 3725–3732.

138. Z. Zhang, Z. Suo and J. He (2005) "Saturated voids in interconnect lines due to thermal strains and electromigration," Journal of Applied Physics, vol. 98, 074501.

139. H. J. Frost and M. F. Ashby (1982) Deformation Mechanism Maps Pergamon Press, Oxford.

140. D. Ang and R. V. Ramanujan (2006) "Hydrostatic stress and hydrostatic stress gradients in passivated copper interconnects." Materials Science and Engineering A, vol. 423, pp. 157–165.

141. J. Zhang, J. Y. Zhang, G. Liu, Y. Zhao and J. Sun (2009) "Competition between dislocation nucleation and void formation as the stress relaxation mechanism in passivated Cu interconnects," Thin Solid Films, vol. 517, pp. 2936–2940.

142. J.-M. Paik, I.-M. Park and Y.-C. Joo (2006) "Effect of grain growth stress and stress gradient on stress-induced voiding in damascene Cu/low-k interconnects for ULSI," Thin Solid Films, vol. 504, pp. 284–287.

143. T. S. Cale, J.-Q. Lu and R. J. Gutmann (2008) "Three-dimensional integration in microelectronics," Chemical Engineering Communications, vol. 195, pp. 847–888.

144. J. Zhang, M. O. Bloomfield, J.-Q. Lu, R. J. Gutmann and T. S. Cale (2006) "Modeling thermal stresses in 3D IC interwafer interconnects," IEEE Transactions on Semiconductor Manufacturing, vol. 19, pp. 437–448.

145. P. De Moor, W. Ruythooren, P. Soussan, B. Swinnen, K. Baert, C. Van Hoof and E. Beyne (2006) "Recent advances in 3D integration at IMEC," in Enabling Technologies for 3-D Integration, Materials Research Society Symposium Proceedings, vol. 970, 0970-Y01-02.

Chapter 5
Electronic Packaging Structures

In conjunction with the advancement of microelectronics, the art and technology of electronic packaging have progressed rapidly over the last several decades. Packaging protects the semiconductor or other electronic systems from external environment during manufacturing, assembly, shipping, handling and normal operation. It also serves to facilitate power delivery and thermal management (to maintain the device temperature below certain limits). A package is thus a combination of dissimilar materials in close contact, leading to deformation mismatches and internal stress buildup as a result of thermal and mechanical loads [1–7].

An example of a modern flip-chip microprocessor package is schematically shown in Fig. 5.1. The silicon chip is attached to an organic package substrate through solder joints. (Note that the metal traces in the bottom region of the chip represent the interconnect lines discussed in Chap. 4.) For enhanced structural integrity the polymer-based underfill material is used in-between the solder joints. The package substrate is in turn soldered to the larger-scale printed circuit board. Heat generated during device operation is conducted away primarily through the thermal interface material/heat spreader assembly toward the heat sink. The miniaturization of chip-level features continues to drive the packaging structure to follow the same trend. The increased feature density and power demand further lead to greater current flow and Joule heat, adding to the already intricate situation.

A great variety of packaging configurations are being used or developed for applications encompassing microelectronics, optoelectronics, MEMS and more traditional electrical components. There is a rich body of practical issues that concerns constrained deformation. In the sections below a select number of topics are treated, with the primary focus being deformation in solder joints. The topics serve as a natural sequel to the previous two chapters – one devoted to films (2D structure) and the other to lines (1D structure). A joint may thus be viewed as a zero-dimension structure. A special case study featuring a complicated material layout – not involving solder joints – will be given in Sect. 5.5.

Compared with the feature size of thin films and on-chip interconnect lines addressed in Chaps. 3 and 4, the overall dimensions of individual solder joints are

Y.-L. Shen, *Constrained Deformation of Materials: Devices, Heterogeneous Structures and Thermo-Mechanical Modeling*, DOI 10.1007/978-1-4419-6312-3_5,
© Springer Science+Business Media, LLC 2010

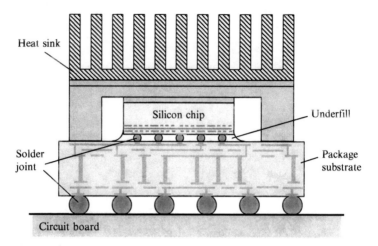

Fig. 5.1 Schematic of a microprocessor package (drawing not to scale). The silicon chip is attached to the package substrate through solder joints. The structure between the chip and heat sink is the thermal interface material/heat spreader assembly [1]

much greater – on the order of 50–500 μm in contemporary devices and can be even larger in other applications or test vehicles. We nevertheless consider these joints as "small" structures in this book. Solder joints are indeed small compared to most structural engineering components, and the mechanical load imposed on them is indirect – through other externally bonded materials. Therefore they fit perfectly into the present discussion on constrained deformation.

Solder joints in microelectronic packages serve the dual purposes of electrical conduction and mechanical connection. Low melting point, good wetting behavior, low cost, and acceptable chemical characteristics on metallization surfaces are some of the key requirements. Traditionally tin (Sn)–lead (Pb) alloys are the mainstream solder material; in recent years Pb-free solder (mainly Sn-rich alloys) as an environment-friendly replacement has received much attention [8–14]. The long-term structural reliability of solder has always been a major concern. Mechanical stresses in solders are generated mainly due to thermal expansion differences between the components they join, as a consequence of temperature fluctuations caused by the operation environment and Joule heat. These stresses have a predominant shear component. Depending on the geometry and location of the individual joint, strains in solder can reach 10% or above. Material degradation is caused not only by cyclic plastic deformation, but also by creep damage because of the inherent low melting point. Vibration and mechanical shock also cause significant solder deformation under severe strain rate conditions.

Thermo-mechanical deformation and damage of Pb-bearing and Pb-free solders have long been under intensive research [15–70]. In Sects. 5.1 to 5.4 we focus on how constrained deformation evolves in the solder joint and the implications. Attention is again on the continuum-level depiction, not on the microstructural aspects, of the material.

5.1 Quantification of Solder Deformation

The primary form of deformation experienced by solder joints in electronic packages is shear. In laboratory testing, solder is also frequently put into a shear deformation mode. The lap-shear technique has thus been widely employed to evaluate the stress–strain, creep, and thermal and mechanical fatigue behavior of solder [15, 18, 20, 21, 23, 31–33, 39, 41, 48, 50, 61]. The solder is placed between two metallic blocks (termed "substrates" here), which are pulled in opposing directions to provide shear loading. A schematic of the lap-shear geometry is shown in Fig. 5.2a. The average shear stress in the solder can be taken as the applied axial load (in the x-direction) divided by the solder joint (or pad) area in contact with the substrate. As for the shear strain, the common practice is to use the applied displacement divided by the solder thickness, i.e.,

$$\gamma = \frac{\Delta l}{h}, \tag{5.1}$$

following the engineering convention. However, as will be demonstrated below, this representation of shear strain is oversimplified and can lead to dramatic errors. There is a need to quantify this form of constrained deformation.

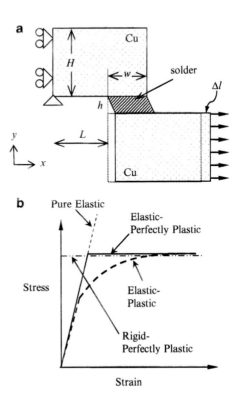

Fig. 5.2 (a) Schematic of the solder/substrate assembly, along with the loading and boundary conditions used in the modeling. (b) Schematics showing the different stress–strain behaviors of solder used for studying the effect of material constitutive response [71]

5.1.1 Lap-Shear Model and Stress Evolution

5.1.1.1 Model Setup

The numerical model is shown in Fig. 5.2a, with the solder joint bonded to two copper substrates [71]. In the simulation horizontal displacements (Δl in the x-direction) are imposed at the far right end of the lower copper substrate. The x-direction motion of the far left edge of the upper copper is forbidden, but movement in the y-direction is allowed except that the lower-left corner of the upper copper is totally fixed. The relative thicknesses of solder and substrate (see below) are chosen such that apparent bending resulting from the shear loading is kept at minimum, so the numerical findings presented in this section are also qualitatively applicable to the double lap-shear configuration where an additional joint/substrate assembly is incorporated to prevent sample rotation [23]. Note that, although the copper substrate considered is relatively thick, it is not a rigid solid and will participate in the deformation process during loading. The calculations are based on the plane strain condition, which effectively simulates the nominal simple shearing mode of the solder. A range of solder thickness (h) from 0.06 to 0.5 mm is considered. Combinations of different solder widths (w) and substrate geometries are also considered for each solder thickness. Unless otherwise specified, the dimensions of the substrate are taken as $H = 6.35$ mm and $L = 1.825$ mm.

In the model Cu is taken to be elastic, with Young's modulus of 114 GPa and Poisson's ratio of 0.31. Unless otherwise stated, the solder is assumed to be elastic-plastic with initial strain hardening. Its input stress–strain response is based on the experimental tensile behavior of bulk Sn-3.5Ag specimens [72, 73]. The solder Young's modulus and Poisson's ratio are 48 GPa and 0.36, respectively, and the initial yield strength is 21.5 MPa. The strain hardening response matches the experimental curve up to 32.3 MPa at a plastic strain of 0.16, beyond which the material is assumed to be perfectly plastic. Although the rate-dependent behavior of solder alloys is already significant at room temperature, we do not take the rate-dependent deformation of solder into account in this set of modeling. This is because the rate effect, which is influenced by solder geometry, will make the comparison of results less straightforward. The rate-dependent behavior will be addressed later.

In order to explore the effects of solder constitutive behavior, three other fictitious solder responses are used in our parametric analysis, as depicted in Fig. 5.2b. Here the uniaxial stress–strain curves are schematically shown, with the "elastic-plastic" response consisting of standard properties described in the previous paragraph. The "elastic-perfectly plastic" model has the same Young's modulus as "elastic-plastic" but with a sharp transition at yielding. The "rigid-perfectly plastic" model is similar to "elastic-perfectly plastic" except the Young's modulus is taken to be infinite. The "pure elastic" model simply assumes a linear elastic behavior without yielding.

5.1.1.2   Evolution of Stress and Deformation Fields

Figure 5.3 shows the modeled shear stress–shear strain curves up to the nominal shear strain of 0.05, for solder joints of width $w=6.35$ mm and three different thicknesses $h=0.06$, 0.12, and 0.5 mm. The shear strain follows the definition in (5.1), which is termed "nominal" shear strain subsequently. The "input" solder shear response, converted from the tensile response used as input in the modeling, is also shown in the figure. Note that the apparent simple shearing operation for the finite-sized specimen is fundamentally different from that of the theoretical pure shear mode. The input response is unique for any solder geometry and nevertheless serves as a reference state for comparison purposes. It is seen that the stress–strain curves are highly dependent on the solder geometry. All three cases show significantly lower shear moduli than the input value, suggesting that the "intrinsic" material behavior is not observed from lap-shear testing. A thinner solder leads to a greater deviation.

Figure 5.4a and b show the contour plots of shear stress, σ_{xy}, in the solder and its vicinity, for the solder thicknesses of 0.06 and 0.5 mm, respectively, under the nominal shear strain of 0.025. The large difference between the two cases is evident. The magnitudes of shear stress inside the thicker solder (Fig. 5.4b) are generally greater than 10 MPa, while in the thinner solder (Fig. 5.4a) only a small area at both ends shows relatively high shear stresses. This is consistent with the overall shear stress in Fig. 5.3 at the same nominal shear strain value (0.025).

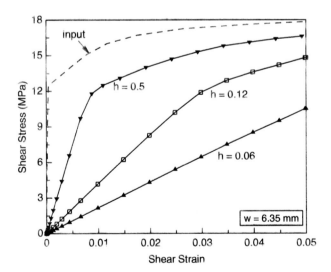

Fig. 5.3 Modeled shear stress–shear strain response of the solders with three different thicknesses: 0.06, 0.12 and 0.50 mm. The solder width is fixed at 6.35 mm. The shear strain is determined based on the "nominal" response. The input shear stress–shear strain response of solder used in the modeling is also shown

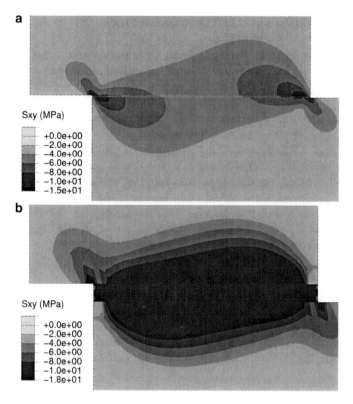

Fig. 5.4 Contour plots of shear stress, σ_{xy}, in the solder and its vicinity, under the nominal shear strain of 0.025. The solder thickness is 0.06 mm in (**a**) and 0.5 mm in (**b**); the solder width is 6.35 mm in both cases. For the purpose of clearly viewing the solder part, only a portion of the copper substrate thickness is included. Due to the chosen definition of shear direction, the shear stress values in the solder are negative

The large dependency of shear response on the solder thickness seen above can be qualitatively understood by the following reasoning. We may consider the entire "copper-solder-copper" assembly as one "specimen," as schematically depicted in Fig. 5.5. It is thus implied that the copper substrate participates in the deformation process as well as the solder (which is true as will be discussed later in subsequent sections). Because the nominal shear strain is $\Delta l/h$, to achieve the same *nominal* strain in solder the imposed displacement needed for the case of a thinner solder is smaller (the upper part of Fig. 5.5). This causes smaller overall deformation in the whole copper-solder-copper assembly. The mechanical load is therefore transmitted through the heavily constrained thin solder without inducing enough shear strain therein. In Fig. 5.3 the thinnest solder (0.06 mm) remains largely elastic within the nominal strain range shown. Conceptually, one may view a very thin solder as an effective "glue" layer. As the solder becomes thicker, it is responsible for an increasing proportion of overall deformation through shearing. (In the case

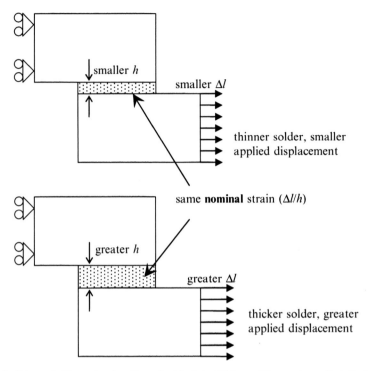

Fig. 5.5 Schematic illustrating the effect of solder joint thickness. The greater applied displacement in the case of a thicker solder causes overall larger deformation in the copper-solder-copper assembly

of a thin solder, the copper substrate carries most deformation which is also smaller due to the smaller applied displacement, as described above.) As a consequence, care must be taken in interpreting experimental results of lap-shear tests. Depending on the solder geometry, the reported shear strain and the actual shear strain solder experiences may differ dramatically. One has to be especially careful when making comparisons of results from different experiments where the specimen geometries may be different. A detailed assessment is given below.

5.1.2 Strain Quantification

The numerical results in this section are presented as shear strain versus applied displacement [71]. Since the real deformation field in the solder is not uniform, the "shear strains" here are calculated by taking the averages of the strain values from all the elements in the solder and thus viewed as "actual" strain values. Figure 5.6 shows the shear strain versus applied displacement curves for two different solder thicknesses: 0.06 and 0.5 mm. Also, the lines of nominal shear strain versus applied

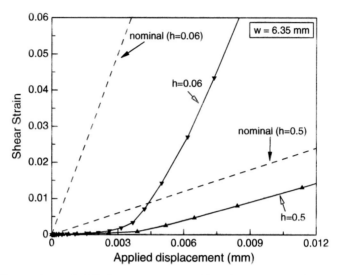

Fig. 5.6 Shear strain–applied displacement response of the solder joints with thicknesses of 0.06 and 0.50 mm. The solder width is fixed at 6.35 mm. The *straight lines* represent the nominal response. The *modeled curves* use the actual shear strains in solder obtained by spatially averaging the strain magnitudes throughout the solder

displacement corresponding to these two thicknesses are included in the figure. (These two nominal lines simply have slopes of inverse solder thickness, which represent the commonly conceived displacement–strain relationship used in practice.) It can be seen that the actual strains in the solder differ significantly from the nominal values. A common feature observed is that, after an initial stage of slow buildup of shear strain in solder, the shear deformation starts to become prominent and the curve tends to be parallel to the nominal one. For a given applied displacement, the thinner solder exhibits a greater deviation in strain whereas the thicker solder is closer to the nominal prediction. This thickness effect is also manifested in the shear stress–shear strain behavior in Fig. 5.3, where the much smaller slope for thin solders is a result of the slower buildup of *actual* shear strain inside the joint.

5.1.2.1   Effect of Solder Constitutive Models

The calculations presented thus far are based on the standard solder response employed as the modeling input, namely the "elastic-plastic" model in Fig. 5.2b. We now compare the shear strain–applied displacement response of all four constitutive models depicted in Fig. 5.2b, as shown in Fig. 5.7. The solder geometry used here is $w=6.35$ mm and $h=0.12$ mm. Again, the "shear strain" represents the averaged elemental values in solder. It is seen that the "elastic-plastic" and "elastic-perfectly plastic" curves in Fig. 5.7 are composed of an initial linear

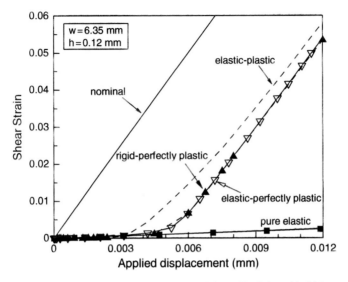

Fig. 5.7 Shear strain–applied displacement response of the solder joint with thickness 0.12 mm and width 6.35 mm. The *straight line* represents the nominal response. Results from the different constitutive behaviors used in the modeling are included

portion and a plastic portion at larger strains. The "pure elastic" response is a straight line whose slope is the same as those of initial portions of the "elastic-plastic" and "elastic-perfectly plastic" curves. The point where these two curves start to deviate significantly from their initial slops can be regarded as the onset of gross yielding. Since the "elastic-perfectly plastic" model has a more extended elastic regime (Fig. 5.2b), plasticity is seen to be more delayed than the "elastic-plastic" model. The "rigid-perfectly plastic" model appears in Fig. 5.7 as very close to the "elastic-perfectly plastic" model. However, an examination of data illustrates that the "rigid-perfectly plastic" model gives rise to zero strain before yielding due to its initial rigidity. Comparing the nominal shear strains with the actual shear strains in solder in Fig. 5.7, it can be concluded that the buildup of shear inside solder is effective only when the solder begins to flow plastically.

5.1.2.2 Effect of Solder Geometry

Figure 5.8 shows the effect of solder geometry on the shear response. Results from four different combinations of solder thickness and width, denoted by "*h/w*" (e.g., 0.12/6.35), are included. The nominal shear strains for the two thicknesses considered (0.06 and 0.12 mm), also included in the figure for reference, are only dependent on the solder thickness. It can be seen that the actual shear response in solder depends not only on the thickness but also on the width of the joint. There is a general trend that a larger aspect ratio (defined as the ratio of thickness/width, or *h/w*) tends to shift the response toward the nominal one. This is evident by comparing

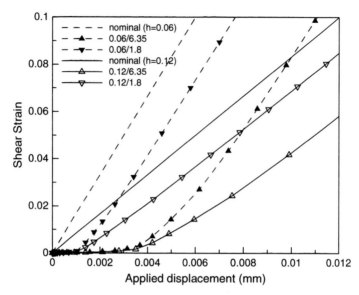

Fig. 5.8 Modeled shear strain–applied displacement response of the solder joints with various geometries (denoted by *h/w*). *Lines without symbols* represent the nominal response

the "0.12/6.35" and the "0.12/1.8" curves to the solid "nominal" line and the "0.06/6.35" and the "0.06/1.8" to the dashed "nominal" line. Among the geometries considered, the thicker and narrower solder exhibits an actual shear strain closer to the nominal case for a fixed applied displacement.

An examination of Fig. 5.8 reveals that the extent of the elastic portion does not depend on the solder thickness. Rather, it is significantly influenced by the width of the solder joint. One can again use a conceptual analysis to rationalize this effect by considering the entire "copper-solder-copper" assembly where both solder and copper can deform. Figure 5.9 shows a schematic of two lap-shear specimens, one with a narrower solder and the other a wider solder, making the entire specimens shorter and longer, respectively. For the same *nominal* shear strain, the same displacement is applied at the far right edges of both assemblies. The induced "overall strain" in the whole assembly containing a wider solder, however, is smaller than the one in the whole assembly containing a narrower solder, since the "overall strain" is simply the applied displacement divided by the overall length. This smaller "overall strain" value in the whole assembly naturally results in a smaller strain in the wider solder, thus prolonging the elastic regime in the longer solder as seen in Fig. 5.8.

5.1.2.3 Comparison with Experiments

The numerical analysis presented above has been verified experimentally [48]. The lap-shear specimens are composed of eutectic Sn-3.5Ag solder and Cu substrates. The substrate thickness is 6.35 mm, and various solder joint thicknesses were used.

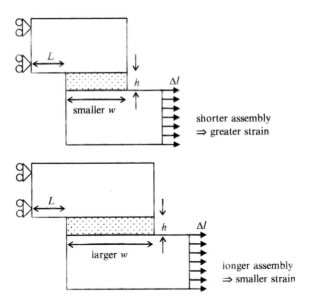

Fig. 5.9 Schematic illustrating the effect of solder width. For the same applied displacement (and thus the nominal strain), the copper-solder-copper assembly with the narrower solder undergoes greater deformation

The nominal strain was measured by an extensometer, 10 mm in span, attached to the Cu substrates. (Note this corresponds to the "standard" substrate geometry used in the modeling above.) Figure 5.10a shows the measured shear stress–shear strain behavior of the solder with various thicknesses. The shear strain measurement is based on the nominal values. Responses for the different thicknesses were obtained using different displacement rates so the nominal shear strain rate is fixed at $10^{-2}\,\mathrm{s}^{-1}$. With increasing solder thickness, the solder exhibits a higher "apparent stiffness." This is consistent with the modeling results presented in Fig. 5.3, thus validating the numerical analyses and relevant discussion. The initial slopes of the curves in Fig. 5.10a are in fact quantitatively similar to those of comparable solder thicknesses in Fig. 5.3. In the fully plastic regime, the experimental curves in Fig. 5.10a converge so that the ultimate shear strength appears to be independent of solder thickness. The apparent experimental yield strength values are higher than those seen in the model. This may be explained by the different intrinsic material behavior for the small-sized solder compared to the bulk material behavior (used for serving as input response in the finite element analysis).

To experimentally measure the *actual* strain the solder carries, a thin line was scribed on the solder surface (along the thickness direction), and the change in displacement of the line was monitored with a traveling microscope [48]. Figure 5.10b shows the measured shear strain as a function of time for the solder thickness of 120 μm. Both the nominal shear strain and the actual solder strain measured by the optical method are included. The test was performed at a constant nominal shear strain rate of $10^{-2}\,\mathrm{s}^{-1}$ so the horizontal axis in Fig. 5.10b also

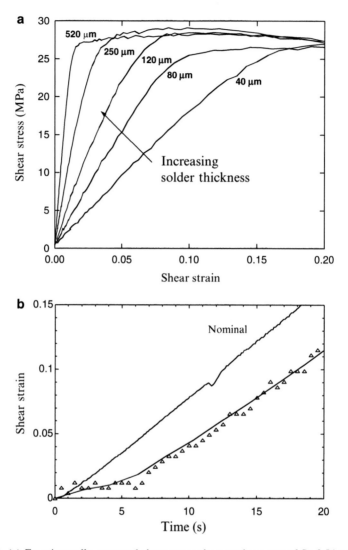

Fig. 5.10 (a) Experimentally measured shear stress–shear strain curves of Sn-3.5Ag solder of different thicknesses. Thicker joints have an "apparent" stiffer response, in consistence with the modeling result. (b) Shear strain versus time in the Sn-3.5Ag solder at a strain rate of about 10^{-2}/s. The actual strain in solder was measured by direct optical sensing

represents the applied displacement. It is evident that, while the nominal strain is linear with time as prescribed by the applied strain rate, the measured actual strain in solder displays the same features as in the numerical results: After an initial stage of slow generation of shear strain in solder, the shear deformation starts to become prominent and the curve tends to be parallel to the nominal one (Figs. 5.6–5.8). The delayed buildup of shear in the solder joint is thus confirmed.

5.1.3   Substrate Geometry and Rigidity

The above discussion on the copper-solder-copper assembly (Figs. 5.5 and 5.9) is based on the notion that the copper substrates can also deform (i.e., they are far from being rigid compared with the solder material). This is indeed the case, as will be evident from the following analysis. Here we consider the effect of substrate geometry. Figure 5.11 shows the modeled shear strain vs. applied displacement plot for four different substrate geometries. The solder geometry is fixed at $h = 0.12$ mm and $w = 6.35$ mm. The nominal line for this solder thickness is also included. Different Cu geometries are denoted by "H/L" (e.g., 6.35/1.825) as defined in Fig. 5.2a. It can be seen in Fig. 5.11 that employing thicker (larger H) and shorter (smaller L) substrates tend to shift the curves towards the nominal response. Thinner and longer substrates prolong the elastic portion of the shear response of the solder joint. Figure 5.11 also includes the results based on simulations treating the substrate as strictly rigid with all other modeling parameters remaining unchanged. The results thus obtained essentially match the nominal response, validating the common practice in lap-shear tests only under this idealized situation. The large geometry dependency revealed in the present analysis can be traced to the deformable nature of the copper substrate. The compliance of the entire solder/substrate assembly contributes to the deformation, which is in fact "absorbed" by the Cu substrate to a significant extent. Although Cu is still elastic, it will not be until the solder becomes plastically free-flowing that the actual shear strain buildup

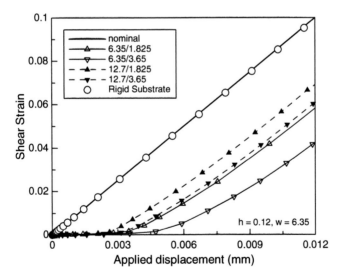

Fig. 5.11 Modeled shear strain–applied displacement response of the solders with various copper substrate geometries (denoted by H/L in mm). The solder geometry is fixed at $h = 0.12$ mm and $w = 6.35$ mm. The nominal response is included. *Circles* following essentially the nominal line are results obtained from treating the substrate material as perfectly rigid

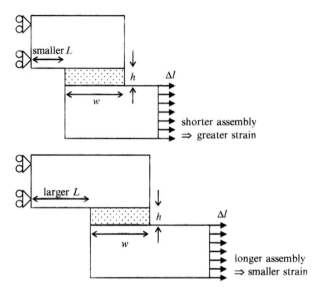

Fig. 5.12 Schematic illustrating the effect of substrate length. For the same applied displacement (and thus the nominal shear strain for solder), the copper-solder-copper assembly with the shorter substrate undergoes greater deformation

in the solder starts to parallel the nominal shear behavior. If a true rigid substrate can be used, the problem of delayed load transfer into solder will not occur.

The effects of substrate geometry can also be realized, again, by the simple conceptual analysis of deformation of the entire copper-solder-copper assembly. Figure 5.12 schematically shows the substrate length effect with a fixed solder geometry. For the same nominal solder strain, the applied displacements for the two cases (shorter and longer substrates) are equivalent. The "overall strain" in the assembly, however, is smaller in the case of a longer substrate pair, which leads to smaller actual strains in solder (or, equivalently, slower buildup of strain in solder). Therefore a greater deviation from the nominal response results (Fig. 5.11). The effect of substrate thickness can be understood by considering the force balance in the assembly, as shown in Fig. 5.13. The applied displacements are the same for the cases of thinner and thicker substrates. The "overall strains" in the two assemblies are therefore equivalent but the pulling force acting on the copper edge for the case of a thicker substrate is greater due to its greater cross section. One can now take the "free body diagram" which is the lower half of the assembly. In the case of a thicker substrate, the greater pulling force is balanced by a greater shear traction acting on the solder, resulting in generally greater shear stresses experienced by the solder. In other words, the buildup of shear in solder in the case of a thinner substrate is not as effective as a thicker substrate. This qualitatively explains the substrate thickness effect.

Figures 5.14a and b show the contour plots of axial stress, σ_{xx}, in the copper-solder-copper assemblies for the copper geometries of (a) $H = 6.35$ mm, $L = 1.825$ mm and (b) $H = 12.7$ mm, $L = 1.825$ mm, respectively. Only a part of each substrate near

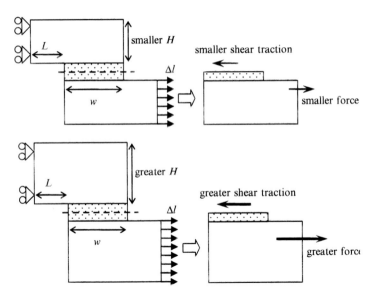

Fig. 5.13 Schematic illustrating the effect of substrate thickness. For the same applied displacement, the case of thicker substrate requires a greater shear traction on solder to balance the pulling force

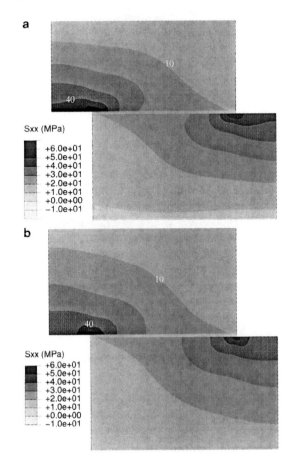

Fig. 5.14 Contour plots of axial stress (σ_{xx}) in the cases of (**a**) $H = 6.35$ mm, $L = 1.825$ mm and (**b**) $H = 12.7$ mm, $L = 1.825$ mm, when the applied nominal strain is 0.025. The solder geometry is $w = 6.35$ mm and $h = 0.12$ mm. Only the parts of the substrates near the solder are shown

the solder is included in the presentation. In both cases the nominal applied shear strain is 0.025. The stresses in the solder are generally small. Although it cannot be specifically seen in Fig. 5.14 due to the choice of contour levels, an examination of data showed that, except near the two edges, the stresses in solder have magnitudes approximately 1 MPa or below, which is significantly smaller than the corresponding shear stresses. In both cases the tensile stresses in copper are quite high near one of the solder edges. However, away from the corner the stress is significantly reduced. This highly localized tensile stress field is a manifest that the overall bending action is kept at minimum when the substrates are sufficiently thick and short. Nevertheless, it is evident that the copper substrate absorbs a considerable amount of deformation. Only if the substrates are rigid can the deformation be entirely borne by solder. For reducing the difference between the nominal and actual shear strains in solder, one needs to use thicker, shorter and stiffer substrates in the lap-shear test.

5.2 Plastic Flow Inside a Solder Joint

The quantification of strain in Sect. 5.1 has focused on the *average* strain value in a solder joint obtained from the modeling. The deformation field inside the metal, however, is non-uniform. In the discussion below we utilize the lap-shear model to examine the evolution of plastic flow. In particular, the plastic deformation field can be correlated with the damage path observed in experiments during thermo-mechanical fatigue. Experiments have shown that solder joint failure can occur at the joint boundary (more specifically, at the interface between the solder material and the intermetallic layer, or inside the intermetallic). This may be explained by the weak interfacial strength or the brittle nature of the intermetallic. However, a commonly observed failure pattern in real-life and laboratory shear-tested solder joints is that cracking appears *close to*, but not at, the interface between the solder alloy and the bonding material [23, 35, 42, 46, 49–51, 74]. This includes both the Sn-Pb and Sn-rich solders, and fracture is entirely within the solder outside of the intermetallic layer. An example is illustrated in Fig. 5.15, which shows a scanning electron microscopy cross-section image of a Sn-3Ag-0.5Cu solder ball after being thermally cycled between −40 and 125°C for 500 cycles [51]. Failure was found to be near the interface between the solder and the Si chip. In the section below the lap-shear model is employed to offer mechanistic insight into this common form of solder damage.

5.2.1 Elastic-Plastic Analysis

The model configuration is identical to that in Fig. 5.2a. The geometry used in the present analysis is defined by $w = 1$ mm, $H = 2.5$ mm and $L = 0.5$ mm. Two different solder thicknesses are considered: $h = 0.125$ and 0.5 mm. The Young's modulus,

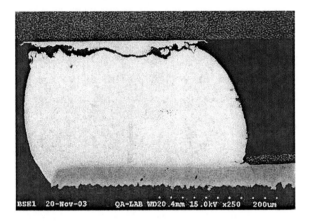

Fig. 5.15 Experimentally observed cracking in a Sn-3Ag-0.5Cu solder ball after thermal cycling between −40 and 125°C for 500 cycles [51]. Reprinted with permission

Poisson's ratio and initial yield strength of the solder are 47 GPa, 0.36 and 20 MPa, respectively. The plastic stress–strain behavior of the solder follows that of Sn-1.0Ag-0.1Cu under a slow uniaxial strain rate of 0.005 s^{-1} [62] up to a peak strength of 36 MPa at the plastic strain of 0.15, beyond which a perfectly plastic response is assumed. Note that only the rate-independent model is considered; the rate-dependent analysis will be discussed in Sect. 5.2.2. Loading is simulated by cyclically imposing nominal solder shear strains between 0 and 0.05. The evolution of equivalent plastic strain field is monitored, because it represents the cumulative irrecoverable deformation which is directly related to the propensity of damage initiation such as microvoid or microcrack formation in ductile metals.

Figure 5.16a and b show the contour plots of equivalent plastic strain in the solder of $h=0.125$ mm, after 3 and 10 full cycles, respectively, between the nominal shear strains of 0 and 0.05. The corresponding plots for the case of $h=0.5$ mm are shown in Fig. 5.17a and b. The plastic strain in the Cu substrates is zero since they remain elastic. It is evident that the plastic flow field in solder is highly non-uniform. Strong plasticity first appears in the four corner regions (with local strain concentration) and tends to propagate into the solder along approximately the 45° direction. However, the dominant deformation mode of the entire joint is horizontal shear, so linking of the plastic localization initiating from the two interface corners tends to occur, leading to a band parallel to each interface. There is no intense plasticity in the mid-thickness sections of the joint, because the material elements there have more freedom to experience rotation, rather than deformation, to help accommodate the applied shear. A comparison between Figs. 5.16a and 5.17a reveals that the strong plastic band develops more easily in the solder with a higher aspect ratio (h/w). After several cycles, the bands become distinct with very high plastic strain values. Note the deformation bands are close to, but not at, the interface between solder and substrates, which can eventually lead to a preferred damage initiation path as observed in many actual tests.

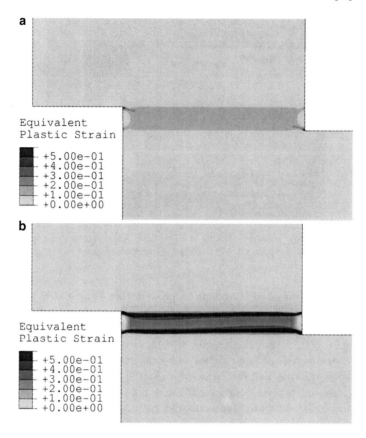

Fig. 5.16 Contour plots of equivalent plastic strain in solder after a deformation history of (**a**) 3 full cycles and (**b**) 10 full cycles between the nominal shear strains of 0 and 0.05. Only a portion of each copper substrate in included in the presentation. The solder geometry is $w=1$ mm and $h=0.125$ mm

Figure 5.18a to c show the contour plots of von Mises effective stress, hydrostatic stress and shear stress σ_{xy}, respectively, in the solder with $h=0.5$ mm, after 10 full cycles of deformation between the nominal shear strains of 0 and 0.05. Although the solder is currently at the overall zero strain state, it can be seen that the stress magnitudes are still quite significant due to the prior plastic deformation history. More importantly, these stress fields do not display any feature that can be corroborated with the damage pattern in actual solders. Therefore the propensity of fracture cannot be attributed to "stress concentration" as claimed by some authors. The equivalent plastic strain field should be used to correlate with the damage initiation path. It is worth pointing out that, if the substrate material is treated as a non-deformable rigid solid, the plastic bands, close to but not at the interface as observed in Figs. 5.16 and 5.17, still exist. Furthermore, if a thin layer of intermetallic between the copper substrate and the solder alloy is included in the model, the same deformation pattern remains [75]. The pattern represents a fundamental deformation feature in an elastic-plastic material under this type of constrained shearing.

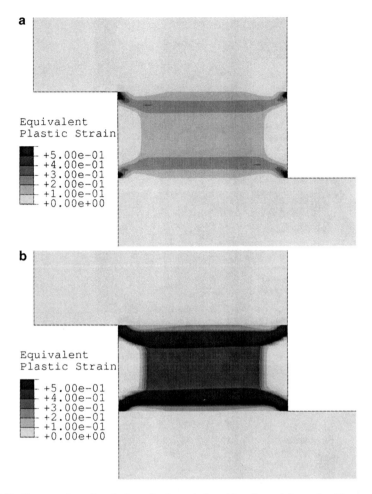

Fig. 5.17 Contour plots of equivalent plastic strain in solder after a deformation history of (**a**) 3 full cycles and (**b**) 10 full cycles between the nominal shear strains of 0 and 0.05. Only a portion of each copper substrate in included in the presentation. The solder geometry is $w=1$ mm and $h=0.5$ mm

5.2.2 Rate-Dependent Model

One factor that can affect the pattern significantly is the loading rate, if the rate-dependent material behavior is incorporated in the model. We now consider the same lap-shear model used in Sect. 5.2.1 with the solder thickness being 0.5 mm. The solder material is taken to be elastic-viscoplastic, with its plastic flow strength beyond the initial yield strength following (2.20),

$$\sigma_e = h\left(\int d\bar{\varepsilon}^p\right) \cdot R\left(\frac{d\bar{\varepsilon}^p}{dt}\right), \tag{5.2}$$

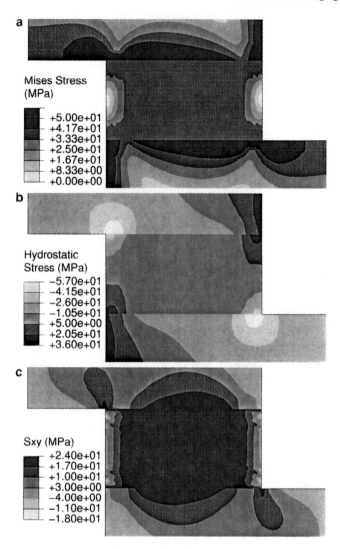

Fig. 5.18 Contour plots of (**a**) von Mises effective stress, (**b**) hydrostatic stress, and (**c**) shear stress σ_{xy} in the solder of $h = 0.5$ mm after 10 full cycles of deformation between the nominal shear strains of 0 and 0.05

where σ_e is the von Mises effective stress, h (a function of plastic strain) is the static plastic stress–strain response, and R (a function of plastic strain rate $\dfrac{d\bar{\varepsilon}^p}{dt}$) defines the ratio of flow stress at nonzero strain rate to the static flow stress where R equals unity. Compared to rate-independent plasticity, this formulation utilizes the scaling parameter R to quantify the strain rate hardening effect. In the illustration below the static stress–strain response of solder is the same as in Sect. 5.2.1.

The R function is based on the experimental measurement of Sn-1.0Ag-0.1Cu alloy [62]: the R values are taken as 1.0, 1.9, 2.4, 2.8, 3.1, 3.4 and 3.5 at the plastic strain rates of, respectively, 0.005, 0.5, 6, 50, 100, 200 and 300 s^{-1}.

Three different applied nominal shear strain rates for the lap-shear analysis, namely 0.01, 1.0, and 100 s^{-1}, are simulated. Figure 5.19a and b show the contour plots of equivalent plastic strain in the solder after 5 and 10 full cycles, respectively, of deformation between the nominal shear strains of 0 and 0.05 under the nominal shear strain rate of 0.01 s^{-1}. The corresponding plots for the strain rates of 1.0 and 100 s^{-1} are shown in Figs. 5.20 and 5.21, respectively. It can be seen that, when the strain rate is relatively low (Fig. 5.19), the plastic flow pattern is qualitatively similar to that in the rate-independent analysis in Sect. 5.2.1. Distinct plastic bands parallel to the interface are formed. As the strain increases, however, a fundamental change in the plastic strain field occurs. At the strain rate of 1.0 s^{-1} (Fig. 5.20), the concentrated deformation band is no longer clear and the overall plastic strain values become smaller compared to Fig. 5.19. This is due to the strain rate hardening effect

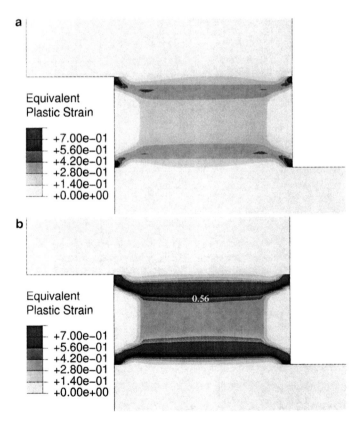

Fig. 5.19 Contour plots of equivalent plastic strain in the solder, after (**a**) 5 and (**b**) 10 full cycles of deformation between the nominal shear strains of 0 and 0.05, under the shear strain rate of $1 \times 10^{-2} \text{s}^{-1}$

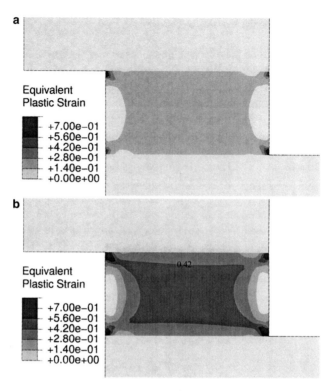

Fig. 5.20 Contour plots of equivalent plastic strain in the solder, after (**a**) 5 and (**b**) 10 full cycles of deformation between the nominal shear strains of 0 and 0.05, under the shear strain rate of 1 s⁻¹

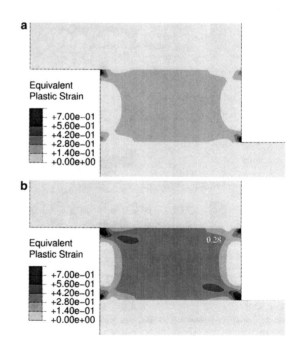

Fig. 5.21 Contour plots of equivalent plastic strain in the solder, after (**a**) 5 and (**b**) 10 full cycles of deformation between the nominal shear strains of 0 and 0.05, under the shear strain rate of $1 \times 10^2 \, \text{s}^{-1}$

in the viscoplastic material so at high strain rates, the initially heavily deformed region becomes hardened, which tends to slow down the accumulation of local plastic strain and diffuse the strain concentration. This trend continues as the applied strain rate becomes even higher, as evidenced in Fig. 5.21 where the strain rate is already at the level encountered in typical drop and impact loading. With continued deformation, failure inside the solder may still be likely due to the accumulation of plasticity, but the lack of a continuous path of plastic localization implies the increasing significance of the interface itself in affecting damage initiation.

5.3   Side Constraint: Effect of Underfill

5.3.1   Shear Loading

In real-life microelectronic packages the solder joints are often encapsulated by polymeric resins on their sides. This "underfill" normally results in improved mechanical reliability of the joints. The elimination of the free side surface can alter the deformation pattern inside the solder, as will be illustrated below. Figure 5.22 shows the model geometry used in the analysis. Although the model still utilizes a simplified 2D scheme, the solder joint takes a more realistic ball shape as in actual packaged devices. The underfill is sandwiched between the upper and lower substrates outside of the solder joint. The entire modeling domain has dimensions of 1×1 mm, with the solder joint of $w = 0.2$ mm and $h = 0.2$ mm. The bottom boundary

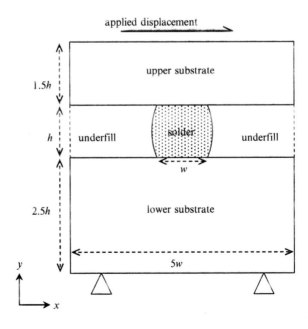

Fig. 5.22 Solder joint model used for studying the effect of side constraint due to the underfill material. In the simulation both w and h are taken to be 0.2 mm. During deformation the bottom boundary is fixed; the top boundary experiences the applied x-displacement but it is not allowed to move in y

of the lower substrate is fixed in space, while the applied displacement along the
x-direction is imposed on the top boundary of the upper substrate during deformation.
The cyclic displacement range between 0 and 0.002 mm is imposed, which results
in a nominal overall shear strain range between 0 and 0.002 (for the entire structure).
The left and right boundaries are not constrained. For simplicity rate-independent
plasticity is used for the solder, which follows the static stress–strain behavior of
Sn-1.0Ag-0.1Cu as in Sect. 5.2.1. The underfill, typically consisting of filled epoxy
resin, is assumed to be elastic with the representative Young's modulus of 7 GPa
and Poisson's ratio of 0.33. For comparison purpose, the numerical result without
the underfill material in the model is also obtained.

5.3.1.1 Solder with Copper Substrates

In the first set of modeling we consider the case with both the upper and lower
substrates to be copper. Figure 5.23a and b show the contour plots of equivalent
plastic strain inside the solder after a loading history of 10 full cycles between
the applied nominal shear displacements of 0 and 0.002 mm, without and with,
respectively, the underfill incorporated. It can be seen in Fig. 5.23a that, without
the side constraint from the underfill, distinct bands of concentrated plasticity
have developed near the solder/substrate interfaces. This deformation pattern
is the same as that in Figs. 5.16 and 5.17, even though the current overall model

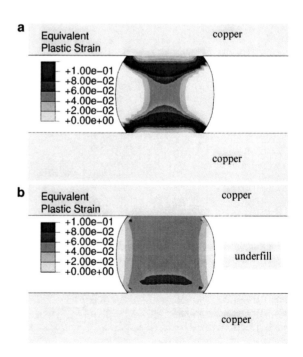

Fig. 5.23 Contour plots
of equivalent plastic strain
in the solder in the models
(**a**) without and (**b**) with the
underfill, after 10 full cycles
of shearing between the
overall nominal shear
displacements of 0 and
0.002 mm. Both the *upper*
and *lower substrates*
are copper

configuration, the shape of the solder joint, and the way shear is applied are different from the treatment before. With the existence of underfill, Fig. 5.23b, however, the banded structure is much less evident. Plastic flow appears to be more uniform, and the overall magnitudes of plastic strain are also significantly reduced. This can explain the reduced propensity of solder damage with the presence of underfill. It is interesting to note that an external polymeric material (much more compliant than the metal) can alter the internal plastic deformation feature inside the metal in a dramatic way.

5.3.1.2   Solder Adjoining Silicon Chip and Circuit Board

We now consider the second set of analysis. The same model geometry as in Fig. 5.22 is used. However, the upper substrate is assumed to be a silicon chip (Young's modulus 130 GPa and Poisson's ratio 0.28) and the lower substrate is assumed to be a printed circuit board (Young's modulus 22 GPa and Poisson's ratio 0.28). Note that the model now becomes a part of simplified flip-chip package structure, and the elastic properties are assumed to be isotropic with approximated values. The same type of cyclic shearing is considered, with the nominal shear displacements between 0 and 0.002 mm applied on the top boundary. Although mechanical loading, not thermal loading, is directly simulated, it is worth pointing out that the present modeling result may also represent, conceptually, the thermal mismatch induced deformation. For instance, if the difference in coefficient of thermal expansion (CTE) between the upper and lower substrates is $15 \times 10^{-6} \, K^{-1}$, a temperature change of 100°C will result in a thermal strain difference of 15×10^{-4} between the two substrates. Consequently, at in-plane locations about 1.4 mm away from the center of an actual chip/board assembly, the overall shear displacement already reaches 0.002 mm.

Figure 5.24a and b show the contour plots of equivalent plastic strain inside the solder without and with, respectively, the underfill after 10 full cycles of deformation. It is seen that, without the side constraint from the underfill (Fig. 5.24a), highly non-uniform plastic deformation occurs. The distinct deformation band, however, appears only near the interface between the solder and the upper substrate. Apparently the stiffness of the substrate material plays an important role in affecting the deformation inside the solder joint. The more compliant lower substrate (in comparison with the upper substrate) tends to deform more easily under the applied shear. In other words, it has a greater tendency to deform *with* the solder so the constraint it imposes on the solder is diminished. It can be deduced from Fig. 5.24a that damage initiation in solder is more likely to take place near the interface with the upper substrate (Si chip). To this end we note that, in Fig. 5.15, the actual crack in the solder ball is also on the Si chip side (the other side is the more compliant organic-based circuit board) [51].

In Fig. 5.24b where the underfill material exists, there is no band with concentrated plasticity inside the solder. The magnitudes of plastic strain are also much reduced compared to Fig. 5.24a. This is the same trend as in the case of Fig. 5.23.

Fig. 5.24 Contour plots of equivalent plastic strain in the solder in the models (**a**) without and (**b**) with the underfill, after 10 full cycles of shearing between the overall nominal shear displacements of 0 and 0.002 mm. In the model the *upper* and *lower substrates* are taken to be silicon and printed circuit board, respectively

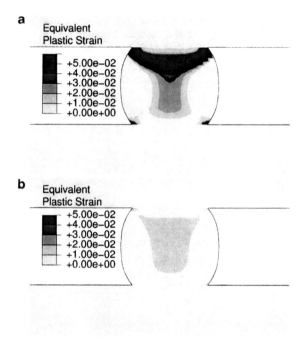

Of course, if one uses more sophisticated material models in the analysis (e.g., rate-dependent response for solder; and/or viscoelastic response for underfill [56]), the deformation pattern will also become a function of the loading rate. A comprehensive understanding of the problem and its implications requires further investigations.

5.3.2 Effect of Superimposed Tension

Up to this point in the present chapter, we consider only the nominal shear deformation of solder joints. Under severe loading conditions such as impact and vibration, bending of the printed circuit board occurs as schematically shown in Fig. 5.25. As a consequence, the solder joints may be forced into tensile or compressive loading in pulsed or cyclic form. In this section we examine the influence of the superimposed tension on the shear deformation. The model configuration is shown in Fig. 5.26. It follows the same geometry as that considered in Sect. 5.3.1, with the upper and lower substrates being the Si chip and printed circuit board, respectively. The only difference is that a displacement in the y-direction can now be applied on the top boundary, generating a pulling action on the solder joint (and the underfill if existent). Although arbitrary histories of tensile and shear displacements may be assigned, we focus on a simple case of in-phase tensile/shear cyclic loading where both the applied displacements in x and y are of the same magnitude, and varied in the range between 0 and 0.002 mm.

Fig. 5.25 Schematic showing solder joint deformation caused by bending of the printed circuit board

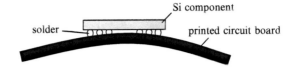

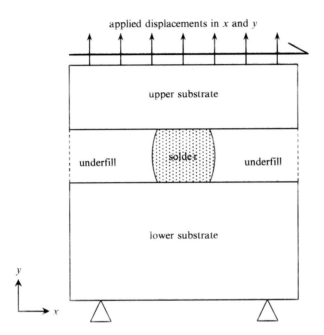

Fig. 5.26 Solder joint model used for simulating the combined cyclic shear and tensile loading. During deformation the bottom boundary is fixed; the top boundary experiences the applied x- and y-displacements

Figures 5.27a and b show the contour plots of equivalent plastic strain in the solder for the cases without and with, respectively, the underfill, after 10 full cycles of combined tensile/shear deformation. It is seen that the superimposed tension drastically changes the plastic flow pattern compared to the simple shear loading in Sect. 5.3.1. The plastic band is aligned in the near 45° direction, apparently the dominant shear direction with respect to the tensile axis. Plasticity is still considerably stronger in the solder joint without the side constraint (Fig. 5.27a), although the banded structure can still be discerned in the case with the underfill (Fig. 5.27b).

From the result above it is realized that damage initiation and failure morphology in the solder can be significantly modified by the existence of normal stresses experienced in actual packages. It will be interesting to conduct a parametric study to quantify the amount of tension (or compression) relative to shear that is required to bring about the change in deformation pattern. The exact form of loading sequence (e.g., cyclic shear with constant compression, monotonically increasing tension with cyclic shear etc.) is also believed to play a role. Further, when the

Fig. 5.27 Contour plots
of equivalent plastic strain
in the solder in the models
(**a**) without and (**b**) with the
underfill, after 10 full cycles
of combined tensile/shear
loading between the overall
nominal tensile/shear
displacements
of 0 and 0.002 mm

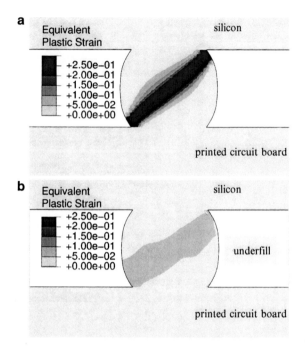

rate-dependent response of solder is taken into account, the effect of loading rate becomes an issue of significance. Some of these problems are given in Sect. 5.6 as suggested project topics.

5.4 Stress Field Around the Solder Joint

The same modeling exercise presented in Sect. 5.3 can now be used to identify the stress values borne by structure. This is a practically important issue because, as a part of the package structure, the Si chip itself is subject to stresses induced by package deformation [6, 76, 77]. Of particular concern is the advanced on-chip interconnects, the integrity of which is dictated by the fragile low-k dielectric material and its weak interface with the metal layer. To tackle the problem one approach is actually to increase the solder joint compliance so stresses transferred to the on-chip structure can be effectively reduced [6]. Despite the fact that the present modeling does not incorporate the fine interconnect structure in the Si chip adjacent to the solder (and underfill), the modeling result still provides a useful picture on the stress field to be expected at these locations.

The evolution of stresses depends on the detailed loading configuration and cyclic deformation history. In the following we make use of the modeling setting in Sect. 5.3, and present selected stress patterns at the end of the first half-cycle, i.e., at the first peak strain (either the nominal shear displacement of 0.002 mm,

Fig. 5.28 Contour plots of
(**a**) von Mises effective stress
and (**b**) maximum principal
stress in the solder joint and
its vicinity, for the case
without the underfill, at the
applied nominal shear
displacement of 0.002 mm
(peak deformation) during
the first loading cycle

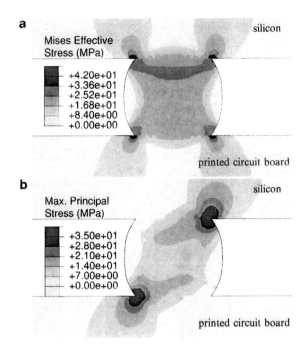

or the same nominal shear displacement superimposed with a nominal tensile displacement of 0.002 mm). The figures below correspond to the following cases: Fig. 5.28 – shear deformation, without underfill; Fig. 5.29 – combined tensile/shear deformation, without underfill; Fig. 5.30 – shear deformation, with underfill; Fig. 5.31 – combined tensile/shear deformation, with underfill. In each of these figures part (a) shows the contour plot of von Mises effective stress and part (b) the maximum principal stress. Note the maximum principal stress, if it is tensile, represents the highest tensile stress experienced by the material locally and may serve as an indication of the propensity of brittle failure initiation.

It can be seen, from Figs. 5.28 and 5.29 (both without the underfill), that the von Mises effective stress in the solder largely displays the banded pattern as the equivalent plastic strain presented in Sect. 5.3, as expected. The largest maximum principal stress in Si, with magnitudes around 40 MPa, appears in localized regions adjacent to the corners of the solder. Approximately the same level of stress is also found in the printed circuit board near the solder. In general, the stresses are moderately higher when the superimposed tension exists (Fig. 5.29). It is worth mentioning that, if the loading continues monotonically to a higher strain (not shown here), the stresses will not become much higher. This is because the upper and lower parts of the structure are connected only by the solder joint, which has already deformed into the plastic regime and thus has limited ability for further strengthening.

If the underfill material exists, different forms of stress distribution are seen. In the case of simple shear loading, Fig. 5.30, the von Mises effective stress and

Fig. 5.29 Contour plots
of (**a**) von Mises effective
stress and (**b**) maximum
principal stress in the solder
joint and its vicinity,
for the case without the
underfill, at the combined
nominal tensile displacement
of 0.002 mm and shear
displacement of 0.002 mm
(peak deformation) during
the first loading cycle

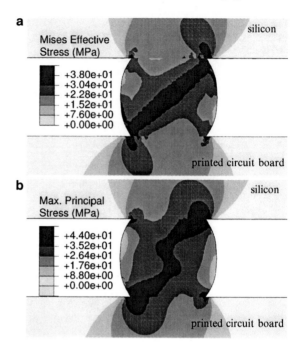

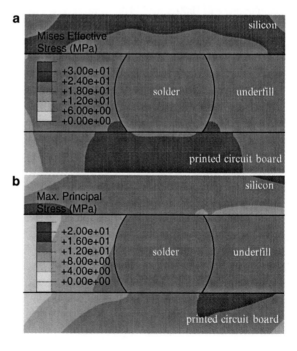

Fig. 5.30 Contour plots of
(**a**) von Mises effective stress
and (**b**) maximum principal
stress in the solder joint
and its vicinity, for the case
with the underfill, at the
applied nominal shear
displacement of 0.002 mm
(peak deformation) during
the first loading cycle

Fig. 5.31 Contour plots of
(**a**) von Mises effective stress
and (**b**) maximum principal
stress in the solder joint and
its vicinity, for the case with
the underfill, at the combined
nominal tensile displacement
of 0.002 mm and shear
displacement of 0.002 mm
(peak deformation) during
the first loading cycle

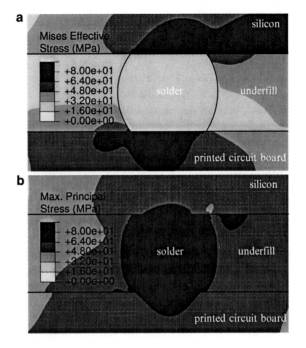

maximum principal stress are generally around or below 30 MPa. The superimposed
tension, however, significantly enhances the stresses, Fig. 5.31. This is apparently
due to the contribution of tensile stress along the *y*-direction. Maximum principal
stresses in excess of 60–80 MPa can occur in the silicon and printed circuit board,
as well as in the solder itself. (The plastic flow strength of solder is well below
these values, so high lateral tensile stresses also exist in solder to "bring down" the
von Mises effective stress to meet the yield criterion (2.15).) If the loading continues
monotonically, the stresses will also increase significantly. The reason is that the
silicon and circuit board are now connected by not just the solder joint but also the
underfill, forming a continuous path across the width to carry the additional shear
and/or tensile stresses.

A comparison between Figs. 5.30 and 5.28 shows that, in simple shear loading,
stresses in some regions of the Si chip can actually be reduced by the presence of
underfill. However, by comparing Figs. 5.31 and 5.29 it is seen that the lower
part of Si chip is under significantly higher stresses in the case with the underfill,
implying the detrimental effect caused by the superimposed tension. The underfill
itself may also experience high stresses and is therefore susceptible to delamination
damage. We conclude by noting that the quantitative stress and strain fields in the
lower part of Si, obtained from the package-level modeling, may be used as the
loading/boundary conditions in on-chip interconnect simulation (such as the models
in Figs. 4.9 and 4.16). The deformation features in the metal lines and dielectric
materials, influenced by the packaging induced stresses, can thus be assessed.

5.5 Multi-component Interaction: A Case Study on Transformer Packaging

An electronic package inherently contains multiple components. In the previous sections attention was devoted to deformation on a "local" scale – in the vicinity of a solder joint. In this section we present a case study of "global" analysis, on an encapsulated transformer which involves multiple components with complex geometries. The reliability issue of concern is cracking in the ferrite core, which is well encapsulated and in close contact with various dissimilar materials. The ferrite core, having a low CTE value, is surrounded by materials with much greater CTEs. Upon cooling down from the encapsulation cure temperature, the core is under overall compression in most regions. However, owing to the complex geometry local tensile stresses along certain orientations can be generated, which tends to promote brittle fracture in the ceramic.

A schematic of a quadrant of the transformer assembly is shown in Fig. 5.32, which is also the model used in the 3D finite element analysis [78]. The core is a ferrite ceramic. Polysulfide is used to form caps to cover the exterior of the core, as well as to fill the inner diameter gap between the core halves (some polysulfide squeezes up between the cores and the bobbin, which is included in the model). Also included in the model are the Delrin bobbin and the copper winding. The transformer structure is completely inside the encapsulation which is a mica filled epoxy resin. The initial design of the transformer did not include the silicone coating. The silicone was applied to the surfaces of the ceramic core as a subsequent modification in an attempt to reduce the residual stress. In the present analysis the same mesh is used in the cases with and without the silicone coating. When silicone is not considered, the coating block is replaced by the encapsulation material.

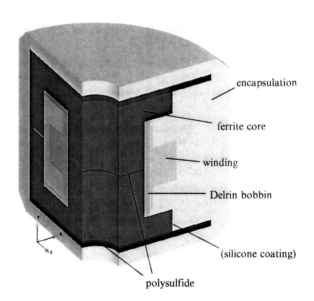

Fig. 5.32 Model of the transformer assembly used in the finite element analysis. Two cases of the model, one with and one without the silicone coating, are considered [78]

encapsulation

ferrite core

winding

Delrin bobbin

(silicone coating)

polysulfide

5.5.1 Elastic Analysis

We first consider the time-independent elastic analysis, where all materials follow linear elasticity. The assembly is assumed to be stress free at the cure temperature (cure shrinkage stresses from polymerization are neglected). The elastic constants and CTE of each material used in the simulation are listed in Table 5.1. Because of the complexity of the winding structure, copper properties are used as an approximation. Thermal stresses are generated in the model by cooling the transformer from the encapsulation cure temperature of 92 to −55°C, where −55 is approximately the minimum temperature environment that this component might be expected to experience during its service duration. In the simulation the mid-horizontal plane and the vertical boundary planes remain planar during deformation because of symmetry. Although the CTE and elastic modulus of the encapsulation are much greater and smaller, respectively, than those of the ferrite core, it is noted that the core is also in direct contact (e.g., the top and bottom parts) with polysulfide which has even greater CTE and lower modulus than the encapsulation. Upon cooling, a stress field far from uniform triaxial compression in ferrite can thus be expected.

Figure 5.33a and b show the contour plots of hydrostatic stress in the encapsulation and maximum principal stress in the ferrite core, respectively, upon cool-down for the case without the silicone coating. It can be seen from Fig. 5.33a that most interior regions in the encapsulation are under a tensile state due to the much greater contraction of the polymer compared to the ceramic. One might expect, from intuition, that embedding the ferrite core in high CTE polymers would result in compressive stresses in the ferrite. However, Fig. 5.33b reveals that significant tensile stresses are generated locally in the ceramic along certain orientations. The main cause of the tensile stresses is the highly uneven contraction of the surrounding materials along with the complex geometry of the structure. For instance, the top and bottom surfaces of the core have a strong tendency to be pulled radially inward due to the neighboring polysulfide and encapsulation. Bending of the cylindrical wall of the core thus induces significant tensile stresses along the ring-shaped region. Another geometrical feature contributing to the tensile stress is the windows at the outer ceramic walls through which wires are passed. The windows result in large volumes of the encapsulation in contact with outer ceramic walls on two sides

Table 5.1 Elastic constants and coefficients of thermal expansion (CTE) used in the modeling of encapsulated transformer

Material	Young's modulus (GPa)	Poisson's ratio	CTE (K^{-1})
Ferrite	161.0	0.25	10.4×10^{-6}
Polysulfide	0.01724	0.40	160.0×10^{-6}
Delrin	2.966	0.35	50.0×10^{-6}
Encapsulation	8.112	0.34	41.0×10^{-6}
Winding	110.0	0.34	18.5×10^{-6}
Silicone	0.01196	0.33	150.0×10^{-6}

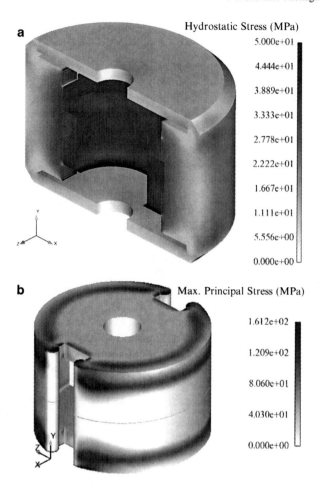

Fig. 5.33 Contour plots of (**a**) hydrostatic stress in the encapsulation and (**b**) maximum principal stress in the ferrite core after cooling to −55°C, in the case of elastic analysis without the silicone coating

of the transformer. This additional encapsulation pulls the top and bottom parts of the vertical walls across the windows together, resulting in additional bending stress in the outer ceramic walls near both windows. Although the polysulfide CTE is much greater than that of the ferrite or encapsulation, the modulus is so low that the polysulfide results in an overall stress reduction. (The polysulfide coating was added for the purpose of reducing stress in the encapsulation. Originally, the exterior encapsulation would crack when applied directly to the ferrite core.) The calculated maximum principal stress in the ferrite is about 161 MPa as seen in Fig. 5.33b. This stress level is significantly higher than the tensile strength of the ferrite core, which is approximately 100 MPa. While this does not necessarily imply that core fracture will occur (due to the different loading configurations and the variability of nominal strength for brittle materials), it is possible that failure at the microscopic scale, such as microcracking, may be initiated during the first cooling process. It is noted that the experimentally observed actual cracking, after further temperature excursions and aging, occurred at the site of highest stressing in Fig. 5.33b [78].

Figure 5.34a and b show the contour plots of hydrostatic stress in the encapsulation and maximum principal stress in the ferrite core, respectively, upon cool-down to −55°C, for the case with a thin silicone coating at the interface where ferrite is in direct contact with the encapsulation in the previous case. From Fig. 5.34a it is observed that the tensile stress state in the interior regions of encapsulation is significantly reduced. Figure 5.34b indicates that the inclusion of the silicone

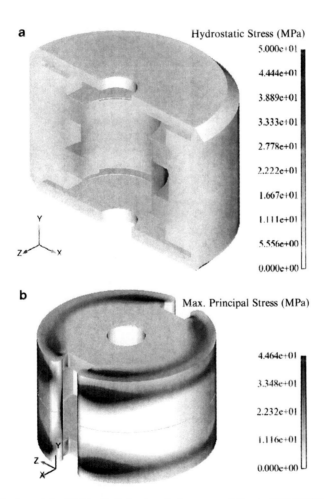

Fig. 5.34 Contour plots of (**a**) hydrostatic stress in the encapsulation and (**b**) maximum principal stress in the ferrite core after cooling to −55°C, in the case of elastic analysis with the silicone coating

coating results in a maximum principal stress of about 44.6 MPa, which is much reduced from the case without the silicone coating in Fig. 5.33b. The silicone coating acts as a buffer to separate the high-CTE encapsulation from the low-CTE ferrite. Without the silicone, a significant state of tensile stress is generated inside the transformer due to the highly confined geometry. With the addition of silicone coating, this problem is alleviated, and cracking was not found in the actual transformer.

5.5.2 Nonlinear Viscoelastic Analysis

The numerical modeling in Sect. 5.5.1 assumed time-independent elasticity. However, more accurate simulations, especially in cases involving physical aging, would require some forms of time-dependent constitutive response in the model. We now present an analysis with the nonlinear viscoelastic response of the encapsulation accounted for, to assess the significance of a prolonged hold at room temperature (physical aging). The model is based on a nonlinear thermo-viscoelastic constitutive equation that has rigorous thermodynamic basis and is capable of predicting a wide variety of mechanical and enthalpic behavior [79–82]. Due to its mathematical complexity the theoretical foundation is not presented here. Rather, we focus on the application of this material model to the same transformer problem as in Sect. 5.5.1. Representative thermo-mechanical properties of the encapsulation resulting from the viscoelastic model are listed in Table 5.2. All the other materials are still assumed to be elastic, with their properties listed in Table 5.1. The simulation first involves cooling of the structure from 105°C to room temperature. A thermal cycle is then included, with cooling to −55°C in 30 min, heating to 105°C in 60 min, and then cooling back to room temperature in 30 min. The transformer is then allowed to age for a period of 4 years. Following aging, the transformer is again subject to the same thermal cycle.

For the case without the silicone coating, the stress contours in the ferrite core after first cooling to −55°C appear qualitatively identical to that in Fig. 5.33b. A plot depicting the simulated maximum principal stress versus temperature at an element in the region of highest tensile stress on the outside diameter of the core is shown in Fig. 5.35. The stress value after the initial cooling to −55°C is about 182 MPa. (Note this value is higher than that in the previous time-independent

Table 5.2 Representative thermo-mechanical properties of the encapsulation resulting from the nonlinear viscoelastic model used in the time-dependent analysis

Property	
Glassy bulk modulus (GPa)	9.3
Glassy shear modulus (GPa)	3.6
Rubbery bulk modulus (GPa)	5.0
Rubbery shear modulus (GPa)	0.062
Glassy linear CTE (K^{-1})	33.3×10^{-6}
Rubbery linear CTE (K^{-1})	116.7×10^{-6}

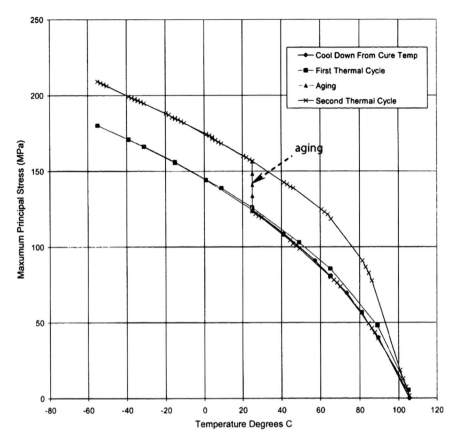

Fig. 5.35 Variation of the maximum principal stress with temperature at an element in the region of highest tensile stress on the outside diameter of the ferrite core, during the entire thermal history (105°C →−55°C → 105°C → 25°C →4 years of aging at 25°C →−55°C → 105°C → 25°C), for the case of viscoelastic analysis without the silicone coating [78]

elastic analysis due to the fundamentally different constitutive model and the slightly different stress-free temperature.) As seen in the figure, at the end of the first thermal cycle (at 25°C), the stress does not differ from the value after initial cooling to the same temperature. The most notable feature, however, is that after the first thermal cycle, the stress increases by 27% (from about 124 to 157 MPa) over the 4-year aging period at room temperature. This may be counterintuitive because one normally conceives that a viscoelastic response results in stress relaxation over time under a fixed kinematic constraint so overall internal stresses in the assembly should be reduced. The main reason for the stress enhancement during aging is the volumetric relaxation exhibited by the encapsulation captured by the nonlinear viscoelastic model. This is confirmed by examining the simulated volumetric strains in representative interior elements in the encapsulation during the thermal cycles and aging. The elements showed notable volume contraction during the isothermal hold period [78]. Effectively, the volume contraction has the same effect of thermal cooling, thus causing the elevation of maximum principal tensile stress in the ferrite core. Figure 5.35 also shows that, during the second thermal cycle (after aging), the maximum principal stress experienced at the outside diameter of the core is 209 MPa, which is a 15% increase from the first thermal cycle. This is an outcome of the stress elevation during aging. Consequently, the failure propensity of the core is increased due to the higher local tensile stress. The viscoelastic modeling therefore offers a rationale for the physical observation: the actual transformer (without the silicone coating) functioned normally during one thermal cycle but experienced failure during a second cycle after a significant aging period [78].

When the silicone coating is included in the model, the tensile stresses in the ceramic core are significantly reduced, with the highest maximum principal stress being about 42 MPa upon first cooling to −55°C. A similar trend of stress enhancement during aging is obtained. The second thermal cycle (after aging) results in a peak maximum principal stress of 44.6 MPa, which is a significant reduction compared to the no-coating design [78]. The improved reliability of the transformer package with the silicone coating can thus be realized from the modeling analysis. This case study demonstrated the merit of systematic computational modeling of constrained deformation, for aiding in our physical insight into the problem and the improvement of component design in real-life applications.

5.6 Projects

1. A systematic study was presented in Sect. 5.1 on the actual strains in the solder joint during the lap-shear testing. Construct a lap-shear model which can accommodate the underfill material on the sides of the joint. Carry out a systematic analysis and re-examine the development of shear strains in the solder with different combinations of solder and substrate geometries as in Sect. 5.1. How different will the results be when the underfill exists?

2. With reference to the rate-dependent deformation of lap-sheared solder joint considered in Sect. 5.2.2, conduct a systematic analysis using the elastic-viscoplastic model featuring a wide range of joint geometries (h/w). Examine the evolution of deformation and stress fields under different applied strain rates. Will the different solder geometries show the same trend as that in Sect. 5.2.2? If different substrate dimensions are used, how will the result be affected?

3. Consider the model in Sect. 5.3.1 (Fig. 5.22) with the upper and lower substrates being the silicon chip and printed circuit board, respectively. Conduct a series of modeling using an elastic-viscoplastic solder material. Focus on the resulting plastic flow pattern in the solder and other stress/strain fields in the vicinity. Study the effects of (1) applied shear rate and (2) underfill material. What if you vary the geometry of the solder joint?

4. With reference to the modeling presented in Sect. 5.3.1, perform similar types of analysis with the viscoelastic properties of the underfill material incorporated. (You may conduct a literature search and decide on a reasonable set of simple linear viscoelastic parameters for use in the model.) What if the rate-dependent elastic-viscoplastic behavior of solder is also included in the model?

5. Construct a solder joint model similar to that in Fig. 5.22 (with the upper and lower substrates being the silicon chip and printed circuit board, respectively). Instead of applying mechanical shearing, try directly modeling thermal cycling with a temperature excursion between, e.g., 125 and −55°C. Think carefully about the boundary conditions. Are you able to achieve sufficiently large shear actions on the solder joint like the mechanical loading case adopted in Sect. 5.3.1? If not, what can be the reason and how would you improve the model configuration to more realistically simulate solder shear due to thermal expansion mismatches as in real microelectronic packages?

6. Consider the modeling analysis in Sect. 5.3.1, with the upper and lower substrates being the silicon chip and printed circuit board, respectively. If the shear displacement is not applied on the top boundary but on the side boundary, will the deformation pattern be different? You may try different combinations of side displacements (pulling or pushing along the x-direction on different sides and/or different substrates). Examine the effects of solder and substrate geometries as well.

7. In Sect. 5.3.2 the effect of superimposed tension on the shear induced plastic flow in solder was explored. The applied nominal tensile displacement was assumed to be in-phase with the shear counterpart and had the same magnitudes. Use the same or a similar model, and expand the analysis to include different relative magnitudes of the applied displacements (e.g., a wide range of the tensile-to-shear displacement ratios from zero to infinity). At what ratio does the shear-dominant flow pattern (plasticity band near the interface) start to disappear? Repeat the analysis for superimposed compression. Can you also think of more complicated cyclic histories of combined tension/compression and shear, and explore their effect on solder deformation?

8. The rising heat generation in modern semiconductor devices has been an issue of great concern. One of the methods for improving thermal management is to incorporate built-in compressive loading (along, for example, the vertical direction in Fig. 5.1) for enhanced thermal conduction efficiency of the thermal interface materials and heat sinks [83]. The superimposed compression is maintained throughout the product life. Conduct a systematic analysis, using the approach in Sects. 5.3 and 5.4, to study the effect of this compression on the deformation and stress fields inside the solder and nearby materials. Discuss the implications with regard to the mechanical reliability.

9. All the simulations of cyclic deformation presented in Sects. 5.2 and 5.3 have used the isotropic hardening model in the plastic constitutive response for solder. Conduct similar types of simulation using the kinematic hardening model instead, and observe the possible differences in the resulting stress and deformation fields. Vary the applied displacement range and cyclic history to explore their influences.

10. In Sect. 5.4 selected stress fields generated in the Si chip due to the applied package deformation were presented. Conduct similar simulations and examine the detailed strain fields in a Si region adjacent to the solder joint. Choose several cases of the strain fields (resulting from different deformation histories) and use them as the loading and boundary conditions in the modeling of on-chip interconnect structure such as in Fig. 4.16. How will the stress and strain fields in the Cu line, barrier layers and dielectric materials be affected?

11. In Sect. 5.1 the nominal shear deformation was considered. Design and carry out a numerical study, using a form of copper-solder-copper assembly, to investigate the nominal *tensile* deformation (e.g., using rod-like specimens with the cylindrical solder joint sandwiched between two Cu cylinders). Vary the solder and copper geometries and examine the evolution of actual stress and strain fields inside the solder, and compare them with the nominal values. Also, extend the analyses to cyclic deformation with various strain amplitudes using both the isotropic and kinematic hardening models. Compare their overall nominal stress–strain responses and discuss the implications, if any.

12. The rate-dependent analysis of solder deformation in Sect. 5.2.2 was based on the elastic-viscoplastic model. Conduct a literature search and identify a creep model for solder (e.g., (2.23) or in other modified forms). Use the creep model to study the effect of loading rate on the deformation pattern, and compare the results with those obtained from the viscoplastic model.

13. Design and carry out a numerical study, using a rate-dependent model (elastic-viscoplastic or creep) for solder, on the *stress relaxation* behavior of solder joints. Impose different strains (shear and/or tensile or compressive, depending on your model) to generate certain states of stress in the solder joint, and observe the decrease in stress over time. Change the solder/substrate geometries to explore the constraint effect. (You may consult experimental studies of stress relaxation in solder in the literature, e.g., [69].)

References

1. V. P. Atluri, R. V. Mahajan, P. R. Patel, D. Mallik, J. Tang, V. S. Wakharkar, G. M. Chrysler, C.-P. Chiu, G. N. Choksi and R. S. Viswanath (2003) "Critical aspects of high-performance microprocessor packaging," MRS Bulletin, vol. 28(1), pp. 21–34.
2. J. Lau, C. P. Wong, J. L. Prince and W. Nakayama (1998) Electronic packaging: design, materials, process, and reliability, McGraw-Hill, New York.
3. D. D. L. Chung, (1995) Materials for electronic packaging, Butterworth-Heinemann, Boston.
4. M. Ohring and J. R. Lloyd (2009) Reliability and Failure of Electronic Materials and Devices, 2nd ed., Academic Press, San Diego.
5. M. C. Shaw (2003) "High-performance packaging of power electronics," MRS Bulletin, vol. 28(1), pp. 41–50.
6. V. Wakharkar and J. C. Matayabas, Jr. (2005) "Future direction and challenges for microelectronics packaging materials," in Proceedings of InterPACK '05, the ASME/Pacific Rim Technical Conference and Exhibition on Integration and Packaging of MEMS, NEMS, and Electronic Systems, paper number: IPACK2005-73374.

7. W. D. van Driel, R. B. R. van Silfhhout and G. Q. Zhang (2009) "On wire failures in microelectronic packages," IEEE Transactions on Device and Materials Reliability, vol. 9, pp. 2–8.

8. W. J. Plumbridge (1996) "Solders in electronics," Journal of Materials Science, vol. 31, pp. 2501–2514.

9. M. Abtew and G. Selvaduray (2000) "Lead-free solders in microelectronics," Materials Science and Engineering R, vol. 27, pp. 95–141.

10. K. Suganuma (2001) "Advances in lead-free electronics soldering," Current Opinion in Solid State & Materials Science, vol. 5, pp. 55–64.

11. K. Zeng and K. N. Tu (2002) "Six cases of reliability study of Pb-free solder joints in electronic packaging technology," Materials Science and Engineering R, vol. 38, pp. 55–105.

12. K. N. Subramanian and J. G. Lee (2003) "Physical metallurgy in lead-free electronic solder development," JOM, vol. 55(5), pp. 26–32.

13. K. N. Tu, A. M. Gusak and M. Li (2003) "Physics and materials challenges for lead-free solders," Journal of Applied Physics, vol. 93, pp. 1335–1353.

14. H. Ma and J. C. Suhling (2009) "A review of mechanical properties of lead-free solders," Journal of Materials Science, vol. 44, pp. 1141–1158.

15. C. Wright (1977) "Effect of solid-state reactions upon solder lap shear strength," IEEE Transactions on Parts, Hybrids, and Packaging, vol. 13, pp. 202–207.

16. W. M. Wolverton (1987) "The mechanisms and kinetics of solder joint degradation," Brazing and Soldering, vol. 13, pp. 33–38.

17. D. Tribula, D. Grivas, D. R. Frear and J. W. Morris, Jr. (1989) "Microstructural observations of thermomechanically deformed solder joints," Welding Research Supplements, October, pp. 404s–409s.

18. D. Frear, D. Grivas and J. W. Morris (1989) "Parameters affecting thermal fatigue behavior of 60Sn-40Pb solder joints," Journal of Electronic Materials, vol. 18, pp. 671–680.

19. E. C. Cutiongco, S. Vaynman, M. E. Fine and D. A. Jeannotte (1990) "Isothermal fatigue of 63Sn-37Pb solder," Journal of Electronic Packaging, vol. 112, pp. 110–114.

20. W. J. Tomlinson and A. Fullylove (1992) "Strength of tin-based soldered joints," Journal of Materials Science, vol. 27, pp. 5777–5782.

21. C. H. Raeder, L. E. Felton, V. A. Tanzi and D. B. Knorr (1994) "The effect of aging on microstructure, room-temperature deformation, and fracture of Sn-Bi/Cu solder joints," Journal of Electronic Materials, vol. 23, pp. 611–617.

22. J. P. Ranieri, F. S. Lauten and D. H. Avery (1995) "Plastic constraint of large aspect ratio solder joints," Journal of Electronic Materials, vol. 24, pp. 1419–1423.

23. D. R. Frear (1996) "The mechanical behavior of interconnect materials for electronic packaging," JOM, vol. 48, pp. 49–53.

24. Z. Guo and H. Conrad (1996) "Effect of microstructure size on deformation kinetics and thermo-mechaical fatigue of 63Sn37Pb solder joints," Journal of Electronic Packaging, vol. 118, pp. 49–54.

25. P. L. Hacke, A. F. Sprecher and H. Conrad (1997) "Microstructure coarsening during thermomechanical fatigue of Pb-Sn solder joints," Journal of Electronic Materials, vol. 26, pp. 774–782.

26. D. R. Liu and Y.-H. Pao (1997) "Fatigue-creep crack propagation path in solder joints under thermal cycling," Journal of Electronic Materials, vol. 26, pp. 1058–1064.

27. P. L. Hacke, Y. Fahmy and H. Conrad (1998) "Phase coarsening and crack growth rate during thermo-mechanical cycling of 63Sn-37Pb solder joints," Journal of Electronic Materials, vol. 27, pp. 941–947.

28. T. Reinikainen, M. Poech, M. Krumm and J. Kivilahti (1998) "A finite-element and experimental analysis of stress distribution in various shear tests for solder joints," Journal of Electronic Packaging, vol. 120, pp. 106–113.

29. J. J. Stephens and D. R. Frear (1999) "Time-dependent deformation behavior of near-eutectic 60Sn-40Pb solder," Metallurgical and Materials Transactions A, vol. 30A, pp. 1301–1313.

30. P. T. Vianco, S. N. Burchett, M. K. Neilsen, J. A. Rejent and D. R. Frear (1999) "Coarsening of the Sn-Pb solder microstructure in constitutive model-based predictions of solder joint thermal mechanical fatigue," Journal of Electronic Materials, vol. 28, pp. 1290–1298.

31. N. M. Poon, C. M. L. Wu, J. K. L. Lai and Y. C. Chan (2000) "Residual shear strength of Sn-Ag and Sn-Bi lead-free SMT joints after thermal shock," IEEE Transactions on Advanced Packaging, vol. 23, pp. 708–714.

32. F. S. Shieu, Z. C. Chang, J. G. Sheen and C. F. Chen (2000) "Microstructure and shear strength of a Au-In microjoint," Intermetallics, vol. 8, pp. 623–627.

33. W. W. Lee, L. T. Nguyen and G. S. Selvaduray (2000) "Solder joint fatigue models: review and applicability to chip scale packages," Microelectronics Reliability, vol. 40, pp. 231–244.

34. Y.-L. Shen, M. C. Abeyta and H. E. Fang (2001) "Microstructural changes in eutectic tin-lead alloy due to severe bending," Materials Science and Engineering A, vol. 308, pp. 288–291.

35. J. H. L. Pang, K. H. Ang, X. Q. Shi and Z. P. Wang (2001) "Mechanical deflection system (MDS) test and methodology for PBGA solder joint reliability," IEEE Transactions on Advanced Packaging, vol. 24, pp. 507–514.

36. J. H. L. Pang, K. H. Tan, X. Shi and Z. P. Wang (2001) "Thermal cycling aging effects on microstructural and mechanical properties of a single PBGA solder joint specimen," IEEE Transactions on Components and Packaging Technology, vol. 24, pp. 10–15.

37. P. L. Tu, Y. C. Chan and K. L. Lai (2001) "Effect of intermetallic compounds on vibration fatigue of μBGA solder joint," IEEE Transactions on Advanced Packaging, vol. 24, pp. 197–205.

38. F. A. Stam and E. Davitt (2001) "Effects of thermomechanical cycling on lead and lead-free (SnPb and SnAgCu) surface mount solder joints," Microelectronics Reliability, vol. 41, pp. 1815–1822.

39. J. H. L. Pang, K. H. Tan, X. Q. Shi and Z. P. Wang (2001) "Microstructure and intermetallic growth effects on shear and fatigue strength of solder joints subjected to thermal cycling aging," Materials Science and Engineering A, vol. 307, pp. 42–50.

40. K. C. R. Abell and Y.-L. Shen (2002) "Deformation induced phase rearrangement in near eutectic tin-lead alloy," Acta Materialia, vol. 50, pp. 3191–3202.

41. J. G. Lee, F. Guo, S. Choi, K. N. Subramanian, T. R. Bieler and J. P. Lucas (2002) "Residual-mechanical behavior of thermomechanically fatigued Sn-Ag based solder joints," Journal of Electronic Materials, vol. 31, pp. 946–952.

42. J. G. Lee, A. Telang, K. N. Subramanian and T. R. Bieler (2002) "Modeling thermomechanical fatigue behavior of Sn-Ag solder joints," Journal of Electronic Materials, vol. 31, pp. 1152–1159.

43. M. Amagai, M. Watanabe, M. Omiya, K. Kishimoto and T. Shibuya (2002) "Mechanical characterization of Sn-Ag-based lead-free solders," Microelectronics Reliability, vol. 42, pp. 951–966.

44. J. Zhao (2003) "Effect of aging treatment on fatigue crack growth in eutectic Sn-Pb alloy," Scripta Materialia, vol. 48, pp. 1277–1281.

45. A. Antoniou and A. F. Bastawros (2003) "Deformation characteristics of tin-based solder joints," Journal of Materials Research, vol. 18, pp. 2304–2309.

46. X. W. Liu and W. J. Plumbridge (2003) "Damage produced in model solder (Sn-37Pb) joints during thermomechanical cycling," Journal of Electronic Materials, vol. 32, pp. 278–286.

47. C. Kanchanomai, Y. Miyashita, Y. Mutoh and S. L. Mannan (2003) "Influence of frequency on low cycle fatigue behavior of Pb-free solder 96.5Sn-3.5Ag," Materials Science and Engineering A, vol. 345, pp. 90–98.

48. N. Chawla, Y.-L. Shen, X. Deng and E. S. Ege (2004) "An evaluation of the lap-shear test for Sn-rich solder/Cu couples: experiments and simulation," Journal of Electronic Materials, vol. 33, pp. 1589–1595.

49. M. Kerr and N. Chawla (2004) "Creep deformation behavior of Sn-3.5Ag solder/Cu couple at small length scales," Acta Materialia, vol. 52, pp. 4527–4535.

50. Y.-L. Shen, K. C. R. Abell and S. E. Garrett (2004) "Effects of grain boundary sliding on microstructural evolution and damage accumulation in tin-lead alloy," International Journal of Damage Mechanics, vol. 13, pp. 225–240.

51. H.-W. Chiang, J.-Y. Chen, M.-C. Chen, J. C. B. Lee and G. Shiau (2006) "Reliability testing of WLCSP lead-free solder joints," Journal of Electronic Materials, vol. 35, pp. 1032–1040.

52. J. Gong, C. Liu, P. P. Conway and V. V. Silberschmidt (2006) "Modelling of Ag_3Sn coarsening and its effect on creep of Sn-Ag eutectics," Materials Science and Engineering A, vol. 427, pp. 60–68.

53. M. A. Dudek, R. S. Sidhu, N. Chawla and M. Renavikar (2006) "Microstructure and mechanical behavior of novel rare earth-containing Pb-free solders," Journal of Electronic Materials, vol. 35, pp. 2088–2097.

54. T. T. Mattila and J. K. Kivilahti (2006) "Reliability of lead-free interconnections under consecutive thermal and mechanical loadings," Journal of Electronic Materials, vol. 35, pp. 250–256.

55. J. Gong, C. Liu, P. P. Conway and V. V. Silberschmidt (2007) "Micromechanical modeling of SnAgCu solder joint under cyclic loading: effect of grain orientation," Computational Materials Science, vol. 39, pp. 187–197.

56. M.-L. Sham, J.-K. Kim and J.-H. Park (2007) "Numerical analysis of plastic encapsulated electronic package reliability: viscoelastic properties of underfill resin," Computational Materials Science, vol. 40, pp. 81–89.

57. J.-W. Jang, A. P. De Silva, J. E. drye, S. L. Post, N. L. Owens, J.-K. Lin and D. R. Frear (2007) "Failure morphology after drop impact test of ball grid array (BGA) package with lead-free Sn-3.8Ag-0.7Cu and eutectic SnPb solders," IEEE Transactions on Electronics Packaging Manufacturing, vol. 30, pp. 49–53.

58. A. Guedon-Garcia, E. Woirgard and C. Zardini (2008) "Reliability of lead-free BGA assembly: Correlation between accelerated ageing tests and FE simulations," IEEE Transactions on Device and Materials Reliability, vol. 8, pp. 449–454.

59. Y.-S. Lai, T. H. Wong and C.-C. Lee (2008) "Thermal-mechanical coupling analysis for coupled power- and thermal-cycling reliability of board-level electronic packages," IEEE Transactions on Device and Materials Reliability, vol. 8, pp. 122–128.

60. C. Y. Chou, T. Y. Hung, S. Y. Yang, M. C. Yew, W. K. Yang and K. N. Chiang (2008) "Solder joint and trace line failure simulation and experimental validation of fan-out type wafer level packaging subjected to drop impact," Microelectronics Reliability, vol. 48, pp. 1149–1154.

61. S.-J. Jeon, S. Hyun, H.-J. Lee, J.-W. Kim, S.-S. Ha, J.-W. Yoon, S.-B. Jung and H.-J. Lee (2008) "Mechanical reliability evaluation of Sn-37Pb solder joint using high speed lap-shear test," Microelectronic Engineering, vol. 85, pp. 1967–1970.

62. E. H. Wong, C. S. Selvanayagam, S. K. W. Seah, W. D. Van Driel, J. F. J. M. Caers, X. J. Zhao, N. Owens, L. C. Tan, D. R. Frear, M. Leoni, Y.-S. Lai and C.-L. Yeh (2008) "Stress–strain characteristics of tin-based solder alloys for drop-impact modeling," Journal of Electronic Materials, vol. 37, pp. 829–836.

63. P. Lall, D. R. Panchagade, P. Choudhary, S. Gupte and J. C. Suhling (2008) "Failure-envelope approach to modeling shock and vibration survivability of electronic and MEMS packaging," IEEE Transactions on Components and Packaging Technologies, vol. 31, pp. 104–113.

64. N. Bai, X. Chen and Z. Fang (2008) "Effect of strain rate and temperature on the tensile properties of tin-based lead-free solder alloys," Journal of Electronic Materials, vol. 37, pp. 1012–1019.

65. F. Liu, G. Meng, M. Zhao and J. F. Zhao (2009) "Experimental and numerical analysis of BGA lead-free solder joint reliability under board-level drop impact," Microelectronics Reliability, vol. 49, pp. 79–85.

66. A. Zamiri, T. R. Bieler and F. Pourboghrat (2009) "Anisotropic crystal plasticity finite element modeling of the effect of crystal orientation and solder joint geometry on deformation after temperature change," Journal of Electronic Materials, vol. 38, pp. 231–240.

67. X. Lin and L. Luo (2009) "Sub-100 μm SnAg solder bumping technology and the bump reliability," Journal of Electronic Packaging, vol. 131, 011014.

68. X. Ong, S. W. Ho, Y. Y. Ong, L. C. Wai, K. Vaidyanathan, Y. K. Lim, D. Yeo, K. C. Chan, J. B. Tan, D. K. Sohn, L. C. Hsia and Z. Chen (2009)"Underfill selection methodology for fine pitch Cu/low-k FCBGA packages," Microelectronics Reliability, vol. 49, pp. 150–162.

69. P. Zimprich, U. Saeed, B. Weiss and H. Ipser (2009) "Constraining effects of lead-free solder joints during stress relaxation," Journal of Electronic Materials, vol. 38, pp. 392–399.

70. E. H. Wong, S. K. W. Seah, W. D. van Driel, J. F. J. M. Caers, N. Owens and Y.-S. Lai, (2009) "Advances in the drop-impact reliability of solder joints for mobile applications," Microelectronics Reliability, vol. 49, pp. 139–149.

71. Y.-L. Shen, N. Chawla, E. S. Ege and X. Deng (2005) "Deformation analysis of lap-shear testing of solder joints," Acta Materialia, vol. 53, pp. 2633–2642.

72. F. Ochoa, J. J. Williams and N. Chawla (2003) "The effects of cooling rate on microstructure and mechanical behavior of Sn-3.5Ag solder," JOM, vol. 55(6), pp. 56–60.

73. F. Ochoa, J. J. Williams and N. Chawla (2003) "Effects of cooling rate on the microstructure and tensile behavior of a Sn-3.5.%Ag solder," Journal of Electronic Materials, vol. 32, pp. 1414–1420.

74. J. W. Morris, Jr., J. L. Freer Goldstein and Z. Mei (1994) "Microstructural influences on the mechanical properties of solder," in The Mechanics of Solder Alloy Interconnects, edited by D. Frear, H. Morgan, S. Burchett and J. Lau, Van Nostrand Reinhold, New York, pp. 7–41.

75. W. H. Moy and Y.-L. Shen (2007) "On the failure path in shear tested solder joints," Microelectronics Reliability, vol. 47, pp. 1300–1305.

76. G. Wang, P. S. Ho and S. Groothuis (2005) "Chip-packaging interactions: a critical concern for Cu/low k packaging," Microelectronics Reliability, vol. 45, pp. 1079–1093.

77. L. L. Mercado, S.-M. Kuo, C. Goldberg and D. Frear (2003) "Impact of flip-chip packaging on copper/low-k structures," IEEE Transactions on Advanced Packaging, vol. 26, pp. 433–440.

78. M. A. Neidigk and Y.-L. Shen (2009) "Nonlinear viscoelastic finite element analysis of physical aging in encapsulated transformer," Journal of Electronic Packaging, vol. 131, 011003.

79. J. M. Caruthers, D. B. Adolf, R. S. Chambers and P. Shrikhande (2004) "A thermodynamically consistent, nonlinear viscoelastic approach for modeling glassy polymers," Polymer, vol. 45, pp. 4577–4597.

80. D. B. Adolf, R. S. Chambers and J. M. Caruthers (2004) "Extensive validation of a thermodynamically consistent, nonlinear viscoelastic model for glassy polymers," Polymer, vol. 45, pp. 4599–4621.

81. D. B. Adolf, R. S. Chambers, J. Flemming, J. Budzien and J. McCoy (2007) "Potential energy clock model: Justification and challenging predictions," Journal of Rheology, vol. 51, pp. 517–540.

82. D. B. Adolf, R. S. Chambers and M. A. Neidigk (2009) "A simplified potential energy clock model for glassy polymers," Polymer, vol. 50, pp. 4257–4269.

83. K. L. E. Helms and B. Phillips (2005) "The characterization of damage propagation in BGA's on flip-chip electronic packages," in Proceedings of InterPACK '05, the ASME/Pacific Rim Technical Conference and Exhibition on Integration and Packaging of MEMS, NEMS, and Electronic Systems, paper number: IPACK2005-73429.

Chapter 6
Heterogeneous Materials

In previous chapters, attention was devoted to materials which are a part of a "structure" so deformation is *externally* constrained by other component(s) of the structure. In many situations, however, there is a structure *inside* the material itself. Examples include all composite and multiphase materials, which are collectively referred to as "heterogeneous materials" here. Strictly speaking, all materials are heterogeneous if the characteristic length scale of concern is at the microstructural or molecular levels. In the present context we focus only on the kind of heterogeneity that is represented by domains with different and well-defined mechanical properties in a given material. When the material is subject to mechanical or thermal loading, *internally* constrained deformation occurs. Such internal constraint naturally dictates the effective (overall) properties of the heterogeneous material, and may also directly affect the propensity of damage initiation.

In the presentation below we attend to the mechanical and thermal-expansion responses of representative heterogeneous materials. Special emphasis is placed on correlating the internal deformation field with the macroscopic behavior, along with practical implications. Issues of importance to micromechanical modeling and interpretation of experimental results are specifically addressed.

6.1 Effective Elastic Response

Predicting the effective elastic modulus of a composite material, based on the known elastic constants of its constituents, has been a subject of significance in engineering design ever since the advent of modern man-made composites. Analytical expressions or approximations exist for certain commonly encountered forms of internal structure in a composite [1–4]. Perhaps the most straightforward type of phase arrangement in a two-phase heterogeneous material, in terms of geometric simplicity, is a laminated structure with alternating layers of the two constituents. Section 6.1.1 discusses the effective elastic response of this type of composites; more complex forms of phase arrangement will follow subsequently.

Y.-L. Shen, *Constrained Deformation of Materials: Devices, Heterogeneous Structures and Thermo-Mechanical Modeling*, DOI 10.1007/978-1-4419-6312-3_6,

6.1.1 Multilayered Structures

The structure of concern consists of alternating layers of stiff (higher-modulus) and compliant (lower-modulus) materials, as schematically shown in Fig. 6.1. Many micro- and nano-scale thin-film laminates targeted for improved mechanical and/or other functional performances fall into this category [5–26]. It is immediately clear that, even if the individual layers A and B in Fig. 6.1 are isotropic, the overall elastic response of the composite will be direction-dependent. The composite is transversely isotropic, with its elastic modulus along the out-of-plane direction (2) different from that along the in-plane directions (1 and 3). A simple way to calculate the in-plane modulus (E_{11}) and out-of-plane modulus (E_{22}) is to invoke the isostrain condition (Voigt model) and isostress condition (Reuss model), respectively.

$$E_{11} = E_A f_A + E_B f_B, \tag{6.1}$$

$$E_{22} = \left[\frac{f_A}{E_A} + \frac{f_B}{E_B} \right]^{-1} . \tag{6.2}$$

Here the Young's moduli of the individual phases are expressed as E_A and E_B and their volume fractions are f_A and f_B. Note that (6.1) and (6.2) are also used in typical materials textbooks to describe the longitudinal and transverse moduli, respectively, of unidirectional continuous-fiber reinforced composites. It should be noted that the two equations are derived under the one-dimensional (1D) assumption, so care must be taken when directly applying them to the present layered composite. For instance, when the layered composite is subject to loading along the 1-direction, uneven deformation in the 3-direction between layers will occur due to the different Poisson's ratios of materials A and B. This will generate stresses in the 3-direction and, in turn, will affect the stresses in the 1-direction and thus the composite modulus E_{11}. Similarly, when the loading is along the 2-direction, the strains along the same direction in the two materials will not be the same. Unequal lateral deformations between the layers will inevitably occur, which will in turn affect stresses in the 2-direction, and thus, the composite modulus E_{22}. As a consequence, depending on the values of the elastic constants, significant errors may exist if one uses these 1D approximations.

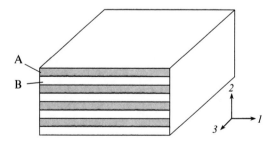

Fig. 6.1 Schematic showing the layered structure with alternating stiff (**a**) and compliant (**b**) materials

6.1.1.1 Numerical Model

Numerical modeling of overall uniaxial loading can be used to accurately determine the composite moduli E_{11} and E_{22}. Figure 6.2a and b show the schematics of compressive loading of the multilayered structure along the 1- and 2-directions, respectively. The composite moduli are calculated from the ratio of stress and strain, obtained from the reaction force and overall displacement in the finite element analysis, along the direction of interest. Although one can construct a true multilayer model for such a purpose, only two representative layers in a small rectangular domain are actually needed, as also shown in Fig. 6.2. Appropriate boundary conditions should be imposed such that a periodic stacking along the out-of-plane direction and an infinite dimension in the in-plane direction are preserved. This is accomplished by setting the top and bottom boundaries to remain horizontal and the side boundaries to remain vertical during deformation. Note in Fig. 6.2a the top boundary is kept horizontal but movement in the 2-direction is allowed, and in Fig. 6.2b the right boundary is kept vertical but movement in the 1-direction is allowed. In addition, the use of the generalized plane strain condition, which allows a constant strain perpendicular to the paper, ensures that the mutual constraint between layers in both the 1- and

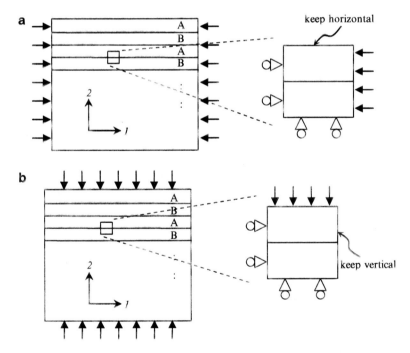

Fig. 6.2 Schematics showing loading along the (**a**) 1- and (**b**) 2-directions for modeling the overall elastic response of the multilayered composite. Due to the fact that all the interfaces remain planar and any arbitrary vertical plane well inside the composite remain vertical during deformation, only a small rectangular domain is needed for the finite element analysis as shown. The applied loading and boundary conditions are also labeled

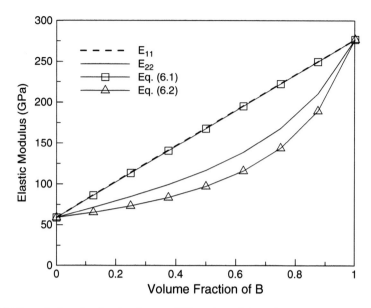

Fig. 6.3 Numerical results showing the overall Young's modulus values along the in-plane direction (E_{11}) and out-of-plane direction (E_{22}) of the A/B multilayers. For comparison the modulus values based on the 1D assumption ((6.1) and (6.2)) are also included

3-directions is properly incorporated. (Alternatively in the case of loading along the 2-direction, Fig. 6.2b, an axisymmetric model can be utilized with the left boundary of the small rectangular domain being the symmetry axis.) The composite response thus simulated is the "true" effective property of the entire multilayer structure, with all three-dimensional features accounted for. In the present example the input material properties are: $E_A = 59$ GPa, $v_A = 0.33$, $E_B = 277$ GPa and $v_B = 0.17$.

Figure 6.3 shows the numerically modeled elastic moduli as a function of volume fraction of material B. For comparison purposes the modulus values given by the 1D approximation ((6.1) and (6.2)) are also included in the figure. It can be seen that the difference between the in-plane modulus (E_{11}) and out-of-plane modulus (E_{22}) of the A/B multilayers is quite large. In general the numerically modeled E_{11} values are close to those given by (6.1), but there is a significant discrepancy between the numerical E_{22} values and those from (6.2). The inaccuracy of applying the 1D approximation to composite modulus of the multilayers is thus illustrated.

6.1.1.2 Overall Elastic Constants

Although the consideration above is limited to the overall moduli in the two directions, we note that the full anisotropic elastic response is actually represented by five elastic constants. For completeness the formulation is given below [27]. First it is convenient to start with the generalized Hooke's law for an orthotropic material system:

$$
\begin{bmatrix}
\varepsilon_{11} \\
\varepsilon_{22} \\
\varepsilon_{33} \\
\gamma_{12} \\
\gamma_{13} \\
\gamma_{23}
\end{bmatrix}
=
\begin{bmatrix}
1/E_{11} & -v_{21}/E_{22} & -v_{31}/E_{33} & 0 & 0 & 0 \\
-v_{12}/E_{11} & 1/E_{22} & -v_{32}/E_{33} & 0 & 0 & 0 \\
-v_{13}/E_{11} & -v_{23}/E_{22} & 1/E_{33} & 0 & 0 & 0 \\
0 & 0 & 0 & 1/G_{12} & 0 & 0 \\
0 & 0 & 0 & 0 & 1/G_{13} & 0 \\
0 & 0 & 0 & 0 & 0 & 1/G_{23}
\end{bmatrix}
\begin{bmatrix}
\sigma_{11} \\
\sigma_{22} \\
\sigma_{33} \\
\sigma_{12} \\
\sigma_{13} \\
\sigma_{23}
\end{bmatrix}
, (6.3)
$$

where ε, γ, σ, E, G and v represent the normal strain, shear strain, stress, Young's modulus, shear modulus and Poisson's ratio, respectively. The coordinate axes are based on those defined in Fig. 6.1. For an orthotropic material the following relations hold true [2]:

$$
\frac{v_{12}}{E_{11}} = \frac{v_{21}}{E_{22}}, \frac{v_{13}}{E_{11}} = \frac{v_{31}}{E_{33}}, \frac{v_{23}}{E_{22}} = \frac{v_{32}}{E_{33}}. \tag{6.4}
$$

Therefore there are a total of nine independent elastic constants (which may be taken as E_{11}, E_{22}, E_{33}, v_{12}, v_{13}, v_{23}, G_{12}, G_{13} and G_{23}). Since the multilayered structure considered here is a special case of the orthotropic material, with transverse isotropy in the 13-plane, therefore

$$
E_{11} = E_{33}, G_{12} = G_{23}, v_{21} = v_{23}, v_{12} = v_{32}, v_{13} = v_{31}, G_{13} = \frac{E_{11}}{2(1+v_{13})}. \tag{6.5}
$$

There are now only five independent elastic constants, which may be chosen as E_{11}, E_{22}, v_{12}, v_{13} and G_{12}. The constants E_{11} and E_{22} can be determined by the numerical approach introduced above. In fact the model in Fig. 6.2a also serves to generate the two Poisson's ratios: v_{12} and v_{13} can be directly calculated from the strain ratios $-\varepsilon_{22}/\varepsilon_{11}$ and $-\varepsilon_{33}/\varepsilon_{11}$, respectively, obtained from the modeling. As for the calculation of effective shear modulus G_{12}, the following analytical expression may be used:

$$
G_{12} = \left[\frac{f_A}{G_A} + \frac{f_B}{G_B} \right]^{-1}, \tag{6.6}
$$

which is based on the equivalent shear stress condition [28, 29]. The five independent elastic constants can thus be completely determined.

6.1.2 Long Fiber Composites

A matrix material reinforced with unidirectional continuous (long) fibers is a common form of engineering composites. The effective elastic modulus of the longitudinal (fiber) direction can be calculated from the rule-of-mixtures (same form as (6.1)) based on the isostrain assumption.

$$E_{longitudinal} = E_M f_M + E_R f_R. \tag{6.7}$$

where the subscripts M and R stand for the matrix and reinforcement phases, respectively. There will still be error due to the Poisson's ratio-induced lateral constraining effect as discussed in Sect. 6.1.1, but the error is typically very small in this case. A greater degree of complexity arises when one considers the elastic modulus in the transverse direction. Below we present a numerical illustration.

Consider fibers with a uniform circular cross section, arranged in a hexagonal packing configuration, inside a matrix as schematically shown in Fig. 6.4a. The fibers are aligned in the 3-direction perpendicular to paper. Due to periodicity and

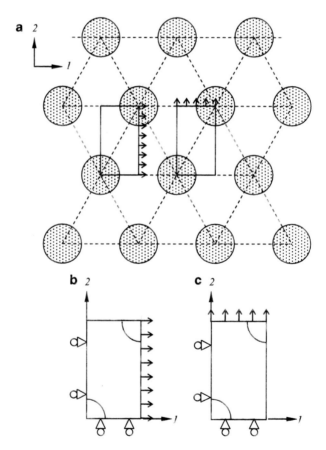

Fig. 6.4 (**a**) Schematic showing the hexagonal array of fibers and the rectangular unit cells used for simulating transverse loading. The long fibers themselves are oriented along the 3-direction. The generalized plane strain unit-cell models for simulating loading along the 1- and 2-directions are shown in (**b**) and (**c**), respectively

symmetry, there is no need to deal with a large computational domain containing many fibers. Instead a unit-cell approach can be adopted, as highlighted by the two rectangular regions in Fig. 6.4a. External tensile loading applied along the 1-direction is schematically shown on the left rectangle, while loading along the 2-direction is shown on the right rectangle. The aspect ratios of the rectangles can be easily determined by geometry to represent the true hexagonal packing. Because of the sixfold symmetry these two loading modes actually represent loading directions only 30° apart (see the overall geometric relationship of the fibers in Fig. 6.4a). A total of 12 loading directions are thus well represented. It is noted that all horizontal sides of the rectangles must be kept horizontal and all vertical sides must be kept vertical during deformation, to preserve the mirror symmetry of the structure. Figure 6.4b and c shows the schematics of the computational domain for the cases of 1- and 2-directional loading, respectively. In both cases the bottom boundary is allowed to move only horizontally and the left boundary is allowed to move only vertically. In addition to the boundary conditions shown, the top boundary in Fig. 6.4b is free to move but has to be constrained to remain straight and horizontal, while the right boundary in Fig. 6.4c is also free to move but has to be constrained to remain straight and vertical. The generalized plane strain condition is imposed, allowing deformation and mutual constraint between the fibers and matrix along the out-of-paper direction. Therefore an actual 3D deformation state caused by transverse loading of the model composite can be obtained.

In this example the composite system is assumed to be an epoxy matrix reinforced with 30 vol.% glass fibers. The Young's modulus and Poisson's ratio of the epoxy are 2.4 GPa and 0.4, respectively, and those of the glass fiber are 72.5 GPa and 0.22, respectively. Loading in both directions is implemented by applying boundary displacements to attain an overall tensile strain of 0.001. Figure 6.5a shows the contour plot of the strain component ε_{11} in the case of loading along the 1-direction; the contour plot of strain component ε_{22} in the case of loading along the 2-direction is shown in Fig. 6.5b. The highly non-uniform strain field is evident in both cases. Due to the higher stiffness of fiber, strains inside the fibers are much smaller compared to those in the epoxy matrix. The strains in the matrix are generally higher in Fig. 6.5a than in 6.5b, especially in regions sandwiched by two fibers along the loading direction (see the global picture of fiber arrangement in Fig. 6.4a). This is because in Fig. 6.5a the matrix region in-between fibers is much shorter along the loading axis, so higher local strains must be generated to compensate for a fixed total applied strain.

Although Fig. 6.5a and b show very different internal strain fields, the effective elastic modulus values, calculated from the ratio of overall stress and strain along the respective loading direction obtained from the finite element modeling, are found to be the same: 4.61 GPa. In fact, the periodic hexagonal array produces theoretical in-plane isotropy for elastic deformation [30, 31]. The present example offers a numerical "check." Or, in reverse thinking, the fact that the same modulus values are recovered from the two models is an indication that the numerical modeling was probably "done right." We also note that, if the 1D isostress formulation were used for the transverse modulus,

Fig. 6.5 Contour plots of
(a) normal strain ε_{11} when the
composite is loaded in
tension along the 1-direction
and (b) normal strain ε_{22}
when the composite is loaded
in tension along the
2-direction. In both cases the
overall tensile strain is 0.001

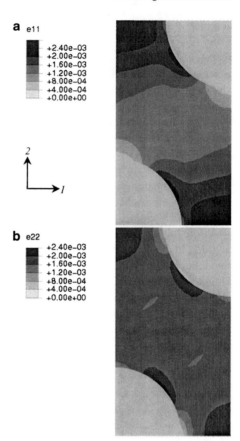

$$E_{transverse} = \left[\frac{f_M}{E_M} + \frac{f_R}{E_R} \right]^{-1}, \qquad (6.8)$$

the composite modulus would be 3.38 GPa. Equation (6.8) can thus lead to a large
error (in this case 27%). Numerical studies on various forms of fiber packing
have been reported [31–34]. In general the transverse modulus is quite sensitive to
the spatial array of the fibers.

6.1.3   A Matrix Containing Discrete Particles

In composites consisting of a matrix with discrete (discontinuous) reinforcement in
all three dimensions, there is no longer a longitudinal direction where the determination of elastic modulus largely follows the isostrain condition. Composites with

a reinforcement shape of short fibers, platelets and particles fall into this category. Particle-reinforced composites can show nominal isotropy in three dimensions, but the internal deformation field caused by mutual constraint of the two phases is very complex.

Various analytical methods have been proposed to estimate the effective elastic modulus of discontinuously reinforced composites, as reviewed in refs. [3, 30, 35–37]. One commonly cited analytical model is the Hashin–Shtrikman upper and lower bounds, which are based on the variational principles of linear elasticity. When applied to an isotropic aggregate, the upper and lower bounds are expressed as [38]

$$K^{upper} = K_R + (1 - f_R) \left[\frac{1}{K_M - K_R} + \frac{3 f_R}{3 K_R + 4 G_R} \right]^{-1} \tag{6.9}$$

$$K^{lower} = K_M + f_R \left[\frac{1}{K_R - K_M} + \frac{3(1 - f_R)}{3 K_M + 4 G_M} \right]^{-1} \tag{6.10}$$

$$G^{upper} = G_R + (1 - f_R) \left[\frac{1}{G_M - G_R} + \frac{6 f_R (K_R + 2 G_R)}{5 G_R (3 K_R + 4 G_R)} \right]^{-1} \tag{6.11}$$

$$G^{lower} = G_M + f_R \left[\frac{1}{G_R - G_M} + \frac{6(1 - f_R)(K_M + 2 G_M)}{5 G_M (3 K_M + 4 G_M)} \right]^{-1}, \tag{6.12}$$

where K and G are the bulk and shear moduli, respectively, and f is the volume fraction of reinforcing particles. The superscripts *upper* and *lower* and the subscripts M and R refer to the upper and lower bound estimates and matrix and reinforcement, respectively. Given the moduli of the matrix and reinforcement, the overall bounds of Young's modulus, E, are then obtained by (6.9)–(6.12) and the relation between E, K, and G,

$$E = \frac{9K}{1 + 3K / G}. \tag{6.13}$$

6.1.3.1 Numerical Model

Numerical finite element modeling can be employed to calculate the effective elastic modulus. Here we focus on the effective Young's modulus under uniaxial loading of the composite. Figure 6.6a shows a schematic of the particle arrangement. The same hexagonal packing as in Sect. 6.1.2 is shown, although in the present case the loading direction is perpendicular to paper and the shaded circles represent cross sections of particles rather than long fibers. The hexagon in Fig. 6.6a shows the true unit cell which contains one particle embedded in the matrix material. To reduce the computations, a common way for approximating the hexagon is to replace it with a circular cell so the 2D axisymmetric condition can be invoked [39–46]. The in-plane elastic isotropy of the hexagonal array also renders this technique

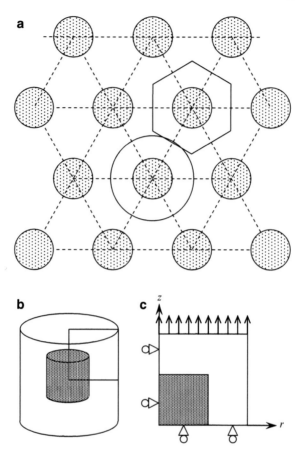

Fig. 6.6 (**a**) Schematic showing a "top view" of the particle arrangement (i.e., viewing direction along the uniaxial loading axis). The hexagon represents a true unit cell in this plane, but it may be approximated by a circular cell so the axisymmetric unit cell, in (**b**), can be used. Part (**c**) shows the computational domain, along with the loading condition in the axisymmetric modeling

a reasonable approximation. Accurate results on the overall mechanical response can be obtained with this approach for moderate particle volume fractions. The axisymmetric unit cell is schematically shown in Fig. 6.6b, where the loading axis of concern is now vertical. The particle therein has a cylindrical shape with its height equal to its diameter (termed "unit cylinder" in the present discussion), although other shapes such as spherical may be used instead. The inserted square represents the computational domain (the unit-cell quadrant) actually needed for the simulation, as highlighted in Fig. 6.6c along with the loading and boundary conditions. The z-axis is the axis of symmetry, and loading is imposed on the top boundary which has to remain horizontal during deformation. In addition to the boundary conditions applied to the bottom and left boundaries, the right boundary is allowed

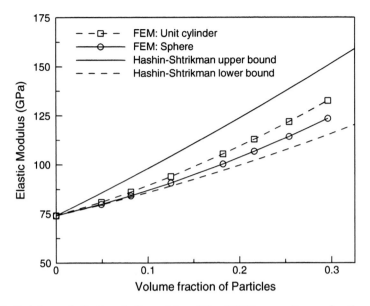

Fig. 6.7 Variations of effective elastic modulus of the Al/SiC composite as a function of SiC particle volume fraction. Finite element modeling (FEM) results on the unit-cylinder and spherical particle shapes, as well as the analytical Hashin–Shtrikman bounds are included

to move but must be constrained to remain vertical, for preserving the perceived alignment of particle in the *r*-direction.

We now present a numerical example using the model in Fig. 6.6c. The matrix material is an aluminum (Al) alloy with Young's modulus 74 GPa and Poisson's ratio 0.33, and the particle is silicon carbide (SiC) with Young's modulus 450 GPa and Poisson's ratio 0.17. Two particle shapes are considered: unit cylinder and sphere. The numerical predictions are shown in Fig. 6.7, where the effective elastic modulus values are plotted as a function of SiC particle volume fraction. For comparison purpose the Hashin–Shtrikman upper and lower bounds are also plotted in the figure. In all cases the effective modulus increases with increasing particle concentration, as expected. It can be seen from the modeling results that the effective modulus is influenced by particle shape. At any fixed particle volume fraction, the unit cylinder particle possesses a greater stiffening effect than the spherical particle, suggesting that the cylinder is a more efficient shape for load transfer (from the compliant matrix to the stiffer reinforcement). Both numerical curves fall within the Hashin–Shtrikman bounds described by (6.9)–(6.13).

It is worth pointing out that the present modeling approach assumes that the particles are aligned in the direction perpendicular to the loading axis. The effective modulus of the composite is also sensitive to the spatial arrangement and size distribution of particles. Examples of the geometric models are given in the sections below, where the effective plastic response of the composite is addressed.

6.2 Effective Plastic Response

The plastic yielding behavior of two-phase aggregates has been a subject of intense interest in the metal matrix composites literature (see reviews in Refs. [29, 47–50]). It has direct implications in the mechanical performance and failure characteristics of this class of composite materials. Numerical modeling is particularly valuable in offering physical insight of constrained plastic flow, with detailed reinforcement geometry and thermo-mechanical history accounted for. We begin the discussion by continuing the particle shape effect in Sect. 6.1.3 and extending the analysis to well into the plastic regime of the matrix material.

6.2.1 Effect of Reinforcement Shape

The model configuration of Fig. 6.6c, with the reinforcement shapes of unit cylinder and sphere, is used. In addition, we consider one case of short fiber-reinforced composite, as schematically shown in Fig. 6.8. The aspect ratio of the fiber, defined to be the ratio of fiber length to fiber diameter, is taken to be 5. The aspect ratio of the whole unit cell is chosen such that the distances from the fiber edge to the cell edge along the r- and z-directions are equivalent (d in Fig. 6.8), in order to allow for uniform fiber spacing in the composite. In all cases the reinforcement volume fraction is assumed to be 12.5%. The elastic properties of the Al alloy matrix and SiC reinforcement used in the modeling are the same as in Sect. 6.1.3. The plastic

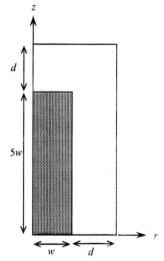

Fig. 6.8 Axisymmetric model used for simulating the composite reinforced with short fibers. The fiber aspect ratio is 5

behavior of the matrix follows that of an age-hardened Al alloy, with an initial yield strength of 390 MPa and linear strain hardening with a slope of 1.59 GPa up to the stress of 520 MPa, beyond which a perfectly plastic response ensues.

The numerically predicted tensile stress–strain curves are plotted in Fig. 6.9 for the various reinforcement shapes. For comparison, the stress–strain curve of the pure matrix is also included. Note the presentation uses both the true stress and true strain, the calculation of which was described in Sect. 2.6. The figure reveals that the tensile response is influenced significantly by reinforcement shape. As expected the short-fiber composite possesses the highest plastic flow stress. The flow stresses for the cases of spherical and unit-cylindrical reinforcement, however, are also quite different. It is noted that the quantitative results in Fig. 6.9 are for a fixed reinforcement volume fraction of 12.5%. The shape effect will become more pronounced if a higher reinforcement concentration is considered. The different degrees of strengthening for the various reinforcement shapes can be explained by examining the local deformation fields.

Figure 6.10a and b show the contour plots of hydrostatic stress and equivalent plastic strain, respectively, for the case of spherical particle-reinforced Al/SiC composite when the applied overall tensile strain is 0.04. The corresponding contour plots for the cases of unit-cylindrical particle and short fiber are shown in Figs. 6.11 and 6.12, respectively. It is seen that the maximum hydrostatic stress in the matrix appears near the top of the reinforcement for all shapes. This is caused by the high

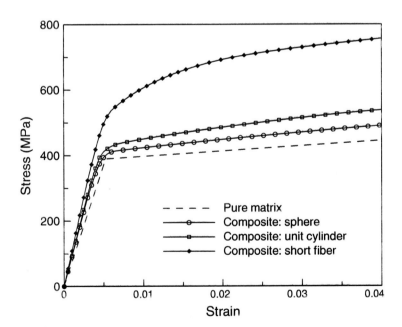

Fig. 6.9 True stress–true strain curves obtained from the finite element modeling, for the Al/SiC composite with spherical, unit-cylindrical and short-fiber reinforcement. For comparison the stress–strain relation used for the matrix material is also included

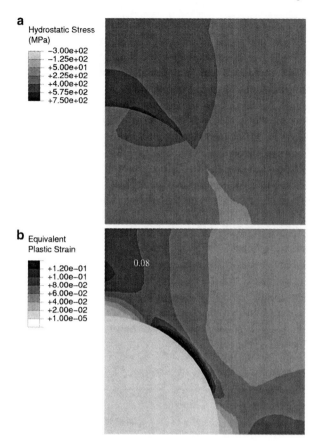

Fig. 6.10 Contour plots of (**a**) hydrostatic stress and (**b**) equivalent plastic strain in the spherical particle-reinforced composite at an applied tensile strain of 0.04

axial tensile stress passing through the elastic reinforcement, together with the lateral tensile stress induced in the same matrix region due to the smaller tendency of lateral contraction of the reinforcement. Although the maximum hydrostatic stress is the highest in the short-fiber composite (Fig. 6.12a), a compressive hydrostatic stress state actually develops in a large region of the matrix around the mid segment of the fiber (i.e., between parallel fibers of adjacent cells).

It can be seen in Figs. 6.10b, 6.11b and 6.12b that intense plastic deformation occurs largely along a direction of about 45° to the tensile axis. In the case of spherical particle, the plastic path appears to be the least disturbed. One can try to picture the periodic arrangement of particles (with other nearby unit cells included) and realize that plasticity tends to follow the widest continuous path in the matrix along the maximum shear (45°) direction. The spherical reinforcement offers the best circumstance among the three, causing easier and more uniform flow which results in the least degree of strengthening. In Fig. 6.11b the plastic flow path is

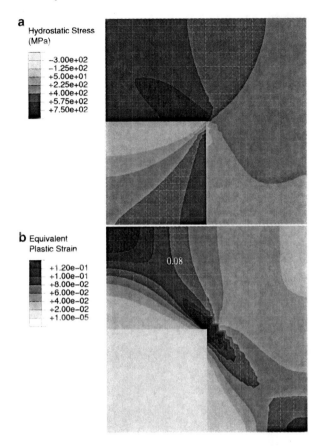

Fig. 6.11 Contour plots of (**a**) hydrostatic stress and (**b**) equivalent plastic strain in the unit-cylindrical particle-reinforced composite at an applied tensile strain of 0.04

interrupted to a larger extent due especially to the sharp corner region of the cylindrical particle. Note the high plastic strain in the "pole" area (above the particle) may be conceived to be caused by the intersection of two plastic bands, one shown in the figure and the other from the adjacent cell quadrant. In the case of short fiber composite, Fig. 6.12b, the matrix flow path is essentially entirely blocked, which forces the 45° band in the upper region to turn toward the vertical side wall of the fiber. Strong load transfer from matrix to fiber and greater overall strengthening can thus be expected.

The mechanism of constrained plastic deformation in reinforced metallic materials was first proposed by Drucker [51–53]. The treatment above illustrated that systematic numerical studies are able to further elucidate how the plastic flow pattern and strengthening behavior are affected by the reinforcement geometry [40, 54]. In the following section we continue to discuss the geometric effect by considering the spatial distribution of reinforcement.

Fig. 6.12 Contour plots of
(**a**) hydrostatic stress and
(**b**) equivalent plastic strain
in the short fiber-reinforced
composite at an applied
tensile strain of 0.04

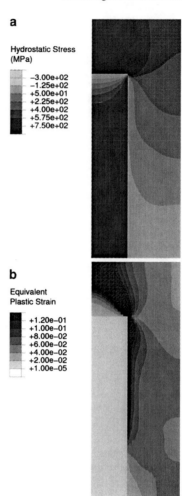

6.2.2 Effect of Reinforcement Distribution

The study of spatial distribution of reinforcement using the unit-cell approach requires multiple particles inside a cell. As a consequence, the use of axisymmetric model is out of the question. Three-dimensional models [55–57] will be needed to attain accurate stress–strain information. Alternatively, larger-scale 3D simulations taking into account more realistic particle distributions, either by design or by mapping the true microstructure inside a composite, can be conducted [58–60]. However, one also recognizes that, while the more realistic composite structure may lead to more accurate overall stress–strain response, a simplified model may be better suited for studying fundamental deformation features. In this section we provide illustrations using the simplified 2D approach.

6.2.2.1 Hexagonal Particle Array

The model with a hexagonal arrangement depicted in Fig. 6.4 can be utilized directly. The tensile loading axis is along the 1- or 2-directions, and the plane strain condition is now imposed with no displacement allowed in the out-of-paper direction. The plane-strain model is able to qualitatively represent the more elaborate 3D calculations [40, 44, 55]. (Note that if the generalized plane strain formulation is used instead, then the modeling becomes one of transverse loading on a long-fiber reinforced composite.) Here we assume the same elastic-plastic properties for the Al alloy matrix and elastic properties for SiC particles as in Sect. 6.2.1. The volume (area) fraction of SiC is fixed at 30%.

Figure 6.13 shows the modeled stress–strain curves loaded along the two directions. For comparison, the stress–strain response of the pure matrix material under the same plane-strain tensile loading is also included in the figure. The two types of loading on the composite results in the same elastic behavior (see also Sect. 6.1.2). However, the plastic response is clearly affected by the relative positions of the

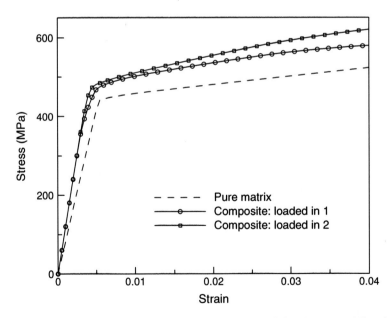

Fig. 6.13 Stress–strain curves obtained from the 2D plane-strain finite element modeling, for the Al/SiC composite with the hexagonal type of SiC particle arrangement shown schematically in Fig. 6.4a. For comparison the plane-strain stress–strain relation of the Al matrix material is also included

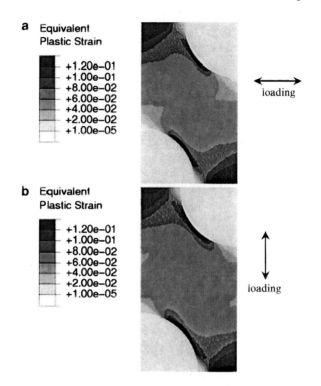

Fig. 6.14 Contour plots of equivalent plastic strain, with tensile loading along the (**a**) 1- and (**b**) 2-directions, when the overall strain is at 0.04. The particle arrangement is shown in Fig. 6.4a

particles with the tensile axis. Loading along the 2-direction results in a higher plastic flow stress.

Figure 6.14a and b show the contour plots of equivalent plastic strain, when the overall tensile strain is at 0.04, for the cases of loading along the 1- and 2-directions, respectively. Strong plasticity is seen to localize in the upper-left and lower-right regions (where different plastic bands intercept), as well as near the edge of the particles. In general, the magnitude of plastic strain is slightly higher in the case of vertical (2) loading, indicating a greater "effort" to bring the ductile matrix to the prescribed macroscopic deformation. From the overall geometric arrangement in Fig. 6.4a, it can be seen that the continuous plastic flow path in the matrix is at an angle of 60° with the loading direction in the case of horizontal loading. As for the vertical loading, the angle is 30°. Upon deformation, the aspect ratios of the rect-angular unit cell are varied in a different manner due to the different loading direc-tions. At the macroscopic strain of 0.04 one can already visually tell this difference in Fig. 6.14. For horizontal loading, Fig. 6.14a, the geometric change is such that the angle between the flow path and loading axis is decreasing from 60° toward 45°. But in vertical loading the flow path is turning further away from the 45° angle (Fig. 6.14b). As a consequence, easier deformation (closer to 45° flow) can be expected in Fig. 6.14a.

6.2.2.2 Effect of Particle Size and Spatial Distribution: Square Array

A different set of investigation on the particle distribution effect is now considered. It is still based on the 2D plane strain calculation, with square-shaped particles making a square type of array. In particular, the particle size and spatial distribution can be treated in a unified way, as depicted in Fig. 6.15. Figure 6.15a represents same-sized particles in a perfectly aligned pattern. In Fig. 6.15b smaller particles are made to exist at the center diagonal position defined by squarely arranged larger particles. The side length of small particles is taken to be one half of that of large particles. For meaningful comparison, the same overall particle concentration is used in all models. In Fig. 6.15c, the small particles in (b) are now "grown" to the same size as the large particles, so the equally sized particles form a staggered arrangement. Parts (a), (b) and (c) are henceforth termed models A, B and C, respectively. The repeated unit structure and the cell quadrant used in the simulation are highlighted in all models. Tensile stretching along the 2-direction is applied on the top boundary of the cell quadrant, while the boundary conditions are the same as in Fig. 6.6c except that the r- and z-axes are now 1- and 2-axes, respectively, in the present plane-strain deformation state. The right boundary is constrained to remain vertical. The volume fraction of SiC particles is kept at 30% in all cases.

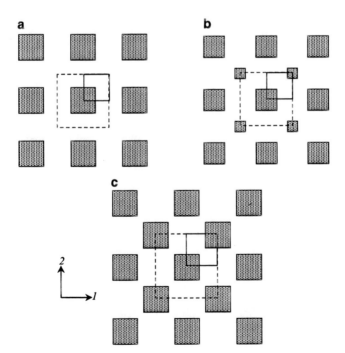

Fig. 6.15 (a) Model A, (b) model B and (c) model C, all with square-shaped reinforcement used for studying the effect of combined spatial and size distributions of particles

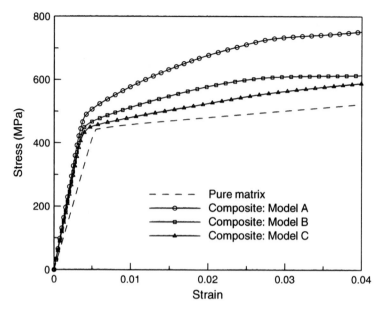

Fig. 6.16 Stress–strain curves obtained from the 2D plane-strain finite element modeling, for the Al/SiC composite with the square-shaped SiC particles. The different particle arrangements are defined in Fig. 6.15. For comparison the plane-strain stress–strain relation of the Al matrix material is also included

Figure 6.16 shows the simulated stress–strain curves for the three models. The case of pure matrix material is also included for comparison. It is evident that particle distribution has a significant effect on the overall composite response. In fact the elastic modulus of the composite is also slightly affected in the present model setting (although it is not easily seen in the figure). Models A and C result in the highest and lowest plastic flow stresses, respectively. It can be seen that model A also shows much greater strengthening compared to the cases in Fig. 6.13 (with the same particle volume fraction), which is caused by the combined effect of shape and spatial distribution of the ceramic particles.

The reasons for the strong particle distribution effect can be understood, again, by examining the plastic flow pattern. Figure 6.17a, b and c show the contour plots of equivalent plastic strain in models A, B and C, respectively, when the applied macroscopic tensile strain is at 0.04. Model A leads to the narrowest plastic band along the 45° direction, with the highest plastic strain values among all three cases. The band in model C is the widest with smallest strain values, signifying that the deformation is more uniform than the other two models. An interesting observation is that, for model B, significant concentration of deformation occurs near the corner of the larger particle. The plastic flow pattern in the matrix is such that there is essentially no plastic deformation in the vicinity of the small particle, which suggests that the small particles in this configuration play essentially no role in the deformation process. This can be appreciated by looking at the overall composite structures in Fig. 6.15: in (b) the continuous paths along the 45° directions in the

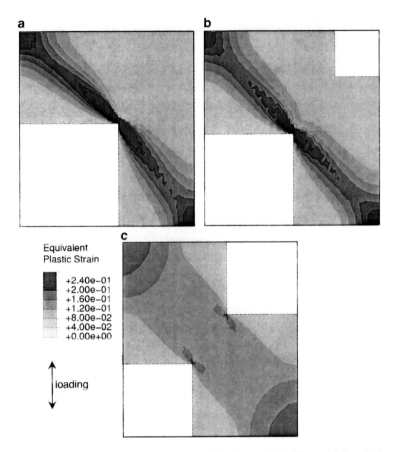

Fig. 6.17 Contour plots of equivalent plastic strain in (**a**) model A, (**b**) model B and (**c**) model C defined in Fig. 6.15, when the applied overall tensile strain is at 0.04

matrix actually do not pass through the small particles. Therefore, plastic deformation is not disturbed by the small particles which make essentially no contribution to the composite strengthening. Model B thus becomes basically a type of model A but with a smaller particle fraction, which results in the lowering of the stress–strain curve. One can perform a numerical exercise by removing the small particles in model B, and examine if the same stress–strain response can be recovered.

If the small particles in model B are now made larger and the larger ones smaller so all particles are of equal size (model C), then the original small particles become relevant in the deformation process. However, in this case the 45° deformation paths in the matrix become uninterrupted and much wider, as evidenced in Figs. 6.15c and 6.17c. The least strengthening is thus obtained.

One may attempt configurations of particle distribution other than those considered in Fig. 6.15. An example is given in Fig. 6.18. Here the large and small particles coexist, but both are arranged in a staggered fashion within its own group. The resulting configuration is an aligned large-small-large type of ordering, which is

Fig. 6.18 A sample configuration of large/small particle
arrangement

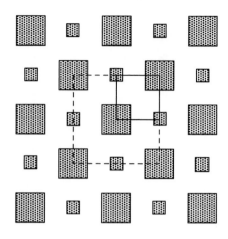

fundamentally different from that in Fig. 6.15b. It can be expected that the small
particles should play an important role in the case of Fig. 6.18, because the 45°
deformation paths in the matrix in-between the larger particles are now interrupted
by the small particles. The small particles are "in the way," thereby contributing to
strengthening of the composite.

6.2.3 Effect of Thermal Residual Stresses

In actual composite materials, internal residual stresses typically exist in the as-
fabricated state. This is mainly due to the thermal contraction mismatch between
the reinforcement and the matrix as the material is cooled from the processing
temperature. In metal matrix composites, the thermal residual stresses may be sufficiently large to yield a large fraction of the matrix. Finite element modeling can
be employed to study the effect of thermal residual stresses on the subsequent
mechanical response of the composite [31, 43, 54, 61–65]. Below we present a case
study using the axisymmetric unit-cell approach as in Sect. 6.2.1.

The model is identical to that in Fig. 6.6c, and we consider both the spherical
and unit-cylindrical SiC particles, with a volume fraction of 12.5%, inside the Al
alloy matrix. The elastic-plastic properties of the two phases are the same as in
Sect. 6.2.1. An isotropic hardening response is assumed for the matrix. To simulate
thermal cooling the coefficients of thermal expansion (CTE) of the constituent
phases are needed. The CTE value of the Al matrix is assumed to be linearly tem-
perature-dependent, with $23.0 \times 10^{-6} \, \text{K}^{-1}$ at 20°C and $30.9 \times 10^{-6} \, \text{K}^{-1}$ at 500°C. The
CTE of SiC is assumed to be independent of temperature with a value of
$4.5 \times 10^{-6} \, \text{K}^{-1}$. The initial stress-free temperature is taken to be 500°C, and the spatially uniform temperature decrease from 500 to 20°C is imposed on the unit-cell

model. During cooling the bottom and left boundaries of the cell quadrant are only allowed to move along the r- and z-directions, respectively. The top and right boundaries are constrained to remain horizontal and vertical, respectively, but otherwise can adjust their positions. Subsequent tensile stretching along the z-direction is applied on the top boundary. Note that after cooling, the model dimensions have changed. As a consequence, when one uses the applied displacement and total pulling force to calculate the true strain and true stress, the after-cooling dimensions should be treated as the initial state.

Figure 6.19a and b show the contour plots of hydrostatic stress in the composites with spherical and unit-cylindrical particles, respectively, after cooling from 500 to 20°C. Due to the facts that the CTE of SiC is much lower and the particles are embedded within the Al alloy matrix, the particles are under high compressive stresses after cooling. It can also be seen that the majority of the matrix is under triaxial tension in

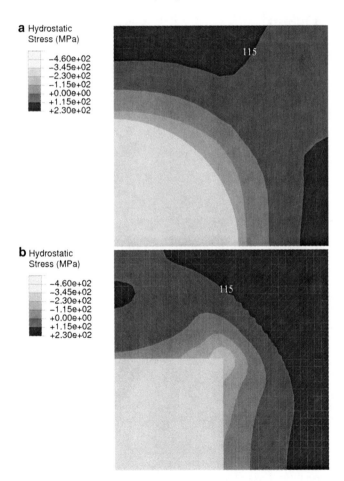

Fig. 6.19 Contour plots of hydrostatic stress in the composites with (**a**) spherical reinforcement and (**b**) unit-cylindrical reinforcement, upon thermal cooling from 500 to 20°C

both cases. Conventional wisdom typically leads to the thinking that the entire metal matrix will be in tension. However, the modeling actually shows a compressive region right adjacent to the ceramic particles. This is caused by the constraining effect not only by the particle but also by the other parts of the matrix. Consider, for instance, the matrix region immediately above the top (flat) end of the particle in Fig. 6.19b. During cooling tensile stress σ_{rr} develops due to the direct CTE mismatch. However, the matrix region occupying the right half of the cell tends to contract along the z-direction, thus imposing compressive σ_{zz} on the matrix immediately above the particle. The net result is that compressive hydrostatic stresses are generated.

Although the reinforcement volume fraction in the present example is only 12.5%, the thermally induced deformation is sufficient to trigger plastic yielding inside the matrix. Figures 6.20a and 6.21a show the contour plots of equivalent

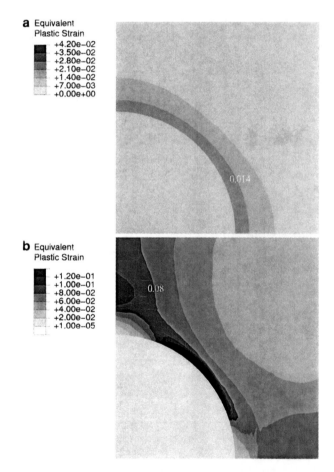

Fig. 6.20 Contour plots of equivalent plastic strain in the composite with spherical reinforcement (**a**) upon cooling from 500 to 20°C and (**b**) after subsequent tensile loading to the macroscopic true strain of 0.04

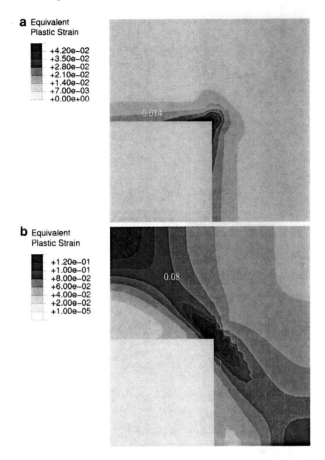

Fig. 6.21 Contour plots of equivalent plastic strain in the composite with unit-cylindrical reinforcement (**a**) upon cooling from 500 to 20°C and (**b**) after subsequent tensile loading to the macroscopic true strain of 0.04

plastic strain after cooling, in the composites with spherical and unit-cylindrical particles, respectively. Plasticity is seen to concentrate around the particles, with the most severely deformed region being close to the corner in the case of unit cylinder (Fig. 6.21a). Figures 6.20b and 6.21b show the corresponding contour plots for the particle shapes of sphere and unit cylinder, respectively, when the composites are subsequently loaded to an overall tensile strain of 0.04. It can be seen that the initial deformation fields have evolved into patterns similar to those of direct mechanical loading without the thermal residual stresses (Figs. 6.10b and 6.11b).

The overall stress–strain curves under the influence of thermal residual stresses are shown in Fig. 6.22. For comparison purpose the curves obtained without the thermal step are also included. It is evident that the presence of residual stresses significantly lowers the elastic modulus of the composite for both the particle shapes. This of course arises from the prior plastic deformation. The stress–strain curves later cross the

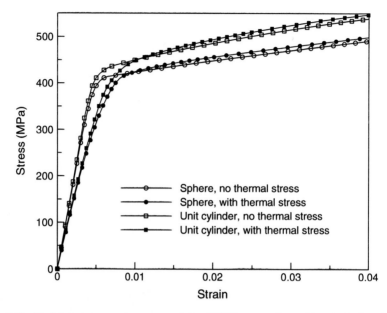

Fig. 6.22 Modeled stress–strain curves of the Al/SiC composites with spherical and unit-cylindrical particles. Both the cases with and without the thermal residual stresses are included for comparison

corresponding curves without thermal residual stresses, and result in higher values of plastic flow stress thereafter. This means that the existence of residual stresses enhances the initial strain hardening. The differences in flow stress beyond the cross-over point persist throughout the plastic regime. This observation points to the fact that the effect of thermal residual stresses can never be eliminated by the development of appreciable plasticity in the matrix during subsequent mechanical loading. The difference in flow stress with and without residual stresses appears to be small in the present case, which is due to the relatively low strain hardening characteristics of the Al matrix in the present models. The flow stress difference will be amplified if stronger matrix strain hardening is in place. It is worth pointing out that, if the matrix material is taken to be perfectly plastic, the presence of thermal residual stresses will only result in transient softening in the elastic and initial plastic ranges, beyond which the stress–strain curve will merge into the perfectly plastic response of the composite (as if the residual stress effect is being "washed out") [61].

6.2.4 Cyclic Response

Attention is now turned to the cyclic stress–strain behavior. In order to adequately model the stabilized cyclic response without encountering elastic shake-down, kinematic hardening is assumed for the Al matrix in this section. We use the

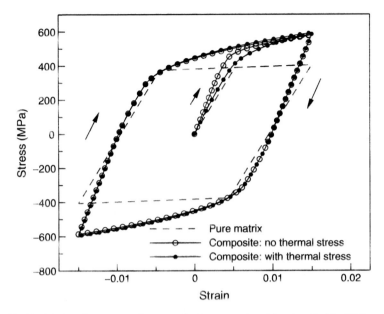

Fig. 6.23 Modeled cyclic stress–strain curves for the composite with unit-cylindrical SiC particles, with and without the existence of thermal residual stresses. The cyclic response of the pure Al alloy matrix is also included. The loading history is 1.25 cycles, with initial loading to the macroscopic strain of 0.015 and then a full cycle between the strains of –0.015 and 0.15

axisymmetric model containing 21.6% volume fraction of SiC having the shape of unit cylinder (Fig. 6.6c). The cyclic history features an initial loading to an overall true tensile strain of 0.015, which is followed by load reversal to the compressive strain of –0.015 and then back to the 0.015 tensile strain. In addition to direct cyclic loading, we also consider the case with thermal residual stresses, obtained by the same method in Sect. 6.2.3 (with a 500 to 20°C cooling step before the mechanical loading).

Figure 6.23 shows the modeled cyclic stress–strain curves of the composite with and without the existence of thermal residual stresses. The corresponding curve for the pure matrix material is also included for comparison. Note that within the strain range considered, the pure matrix only shows a slight extent of strain hardening. A stabilized hysteresis loop is reached after the first cycle. Comparing the curves in the first 1/4 cycle (i.e., up to the first peak tensile strain), one observes similar features as those in Fig. 6.22. The presence of residual stresses results in softer initial response but a crossover of the stress–strain curves later takes place. (The maximum strain chosen in this example is only a little over the crossover point.) It is noticeable that, while the thermal residual stresses result in a smaller apparent elastic modulus due to prior plastic deformation, the "normal" elastic modulus of the composite is recovered upon unloading from the 0.015 peak strain (and reloading from –0.015 strain). Another feature worthy of mention is that, for the pure matrix material, the straight elastic portion of the stress–strain loop has a stress span of 780 MPa (twice the magnitude of initial yield strength, cf. the kinematic hardening model, Sect. 2.3.1). A careful look at the composite loop reveals that, although the composite shows a significant strengthening effect over the matrix, its straight elastic portion is actually smaller than 780 MPa. This arises from the non-uniformity of deformation inside the matrix caused by the particle as seen in previous sections. High *local* effective stress triggers early *local* yielding after the load is reversed, which is reflected in the macroscopic stress–strain response. In other words, constrained cyclic plasticity enhances the Bauschinger effect in the composite.

6.2.5   Explicit Versus Homogenized Two-Phase Structures

Frequently, the two-phase composite microstructure can be homogenized in the modeling to aid in the improvement of computational efficiency and reducing the model complexity. In this section an example of a "three-phase" composite is discussed. The material system is still the same as in the previous sections: an Al alloy matrix reinforced with SiC particles. In actual materials of this type manufactured by normal liquid-infiltration or powder methods, intermetallic inclusion particles inherently exist. The overall concentration of the iron-rich inclusion phase is typically small, but the individual inclusion particles can be greater in size compared to the individual SiC reinforcing particles. The inclusion particles are known to play an essential role in damage initiation [66, 67], so how the local stress and deformation fields evolve in and around the inclusion becomes an issue of significance.

Two models of the same composite are considered here, the cell quadrants of which are depicted in Fig. 6.24. The 2D plane-strain models feature square-shaped SiC particles and inclusion particles. The volume fractions of the SiC phase and inclusion phase are 30% and 5%, respectively. In each cell quadrant the inclusion particle occupies the lower-left corner region. In Fig. 6.24a the region outside the inclusion is a homogeneous material bearing the properties of the Al/30 vol.% SiC two-phase composite. The elastic modulus and plastic response of the Al/SiC mixture are determined based on the experimentally measured stress–strain curve of the composite. Its Poisson's ratio, needed as input for the Al/SiC mixture as well, is obtained from a separate numerical modeling of SiC particle in an Al matrix using the axisymmetric unit-cell approach (as in Fig. 6.6c). Figure 6.24b shows a different configuration with the same inclusion, except that the SiC particles are explicitly included. The number, size and spatial distribution of SiC particles are randomly chosen, but the total SiC fraction is fixed at 30%. It is noted that a quadrant of the representative cell is shown so the inclusion particle is indeed significantly larger than the SiC particles in the model. The models in Fig. 6.24a and b are henceforth termed "homogenized" and "explicit" models, respectively. Tensile loading is applied on the top boundary by way of imposed displacements along the 2-direction. The boundary conditions are identical to those used before in the plane-strain analyses in Sect. 6.2.2. The SiC phase is taken to be elastic with Young's modulus 450 GPa and Poisson's ratio 0.17. The inclusion phase is also assumed to be elastic,

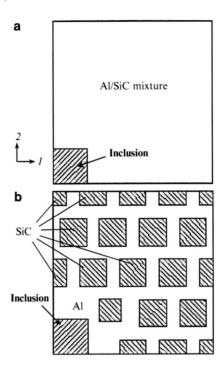

Fig. 6.24 The (**a**) homogenized and (**b**) explicit models of the Al/SiC composite containing the inclusion phase. The square domain represents the cell quadrant used in the 2D plane-strain modeling. The area fractions of SiC particles and the inclusion particle are 30 and 5%, respectively. Tensile loading is applied on the top boundary along the 2-direction [68]

with Young's modulus 260 GPa and Poisson's ratio 0.3. The properties of the Al matrix are based on an 2080 aluminum alloy under the T8 peakaging condition, with Young's modulus 71.8 GPa, Poisson's ratio 0.31, initial yield strength 500 MPa and strain hardening exponent 0.06 [68].

Figure 6.25 shows the simulated stress–strain curves of the composite using the homogenized and explicit models. It is clear that the two models result in nearly identical composite response. If our objective is to obtain the macroscopic stress–strain behavior of this type of three-phase structure, the present result suggests that a homogenized approach would be adequate. However, if local deformation characteristics are sought, the homogenized model will lose its fidelity as illustrated below. Figure 6.26a and b shows the contour plots of equivalent plastic strain in the homogenized and explicit models, respectively, at a macroscopic plastic strain of 0.002. The contrast of the constrained deformation between the two cases is evident. In Fig. 6.26a the deformation field in the Al/SiC mixture is relatively uniform, having a wide and weak plastic strain distribution. Figure 6.26b, however, reveals that the plastic flow field is severely disturbed and localized into narrow bands flowing through the Al matrix following apparently the maximum shear directions. This localized feature, and thus the interaction between the inclusion and SiC particles, cannot be captured

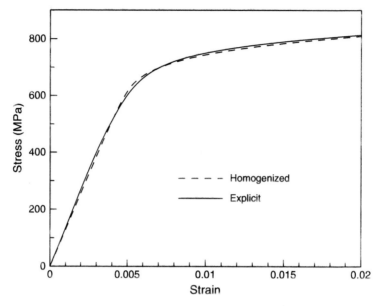

Fig. 6.25 Comparison of the overall stress–strain curves of the composite obtained from the homogenized and explicit models defined in Fig. 6.24

properly by homogenizing the SiC particles and the Al matrix even though the two models yield nearly the same macroscopic response. The different intensities, orientations and widths of the plastic deformation band intersecting the upper-right corner of the inclusion particle in Fig. 6.26a and b no doubt contributes to different stresses carried by the inclusion. This will have direct influences on the failure propensity predicted by the numerical models, a systematic analysis of which can be found in Ref. [68]. We conclude by remarking that, when the micromechanical details are of concern, the composite should be viewed as a "structure," not a simple "material," so a well refined approach (namely the explicit model) should be attempted.

6.3 Effective Thermal Expansion Response

Coefficient of thermal expansion (CTE) is an important physical and mechanical property. As seen in previous chapters, the mismatch in thermal expansion between adjoining materials results in constrained deformation and thermal stresses. These stress and deformation fields exist internally even in a stand-alone heterogeneous material, due to the mismatch of its constituent phases when a temperature change is involved. The effective CTE of a heterogeneous material is thus determined by this mutual constraint at the microscopic level. The numerical unit-cell approach used extensively in the previous sections can be employed to study the effective thermal expansion behavior. In Sects. 6.3.1–6.3.3 below we continue to use the

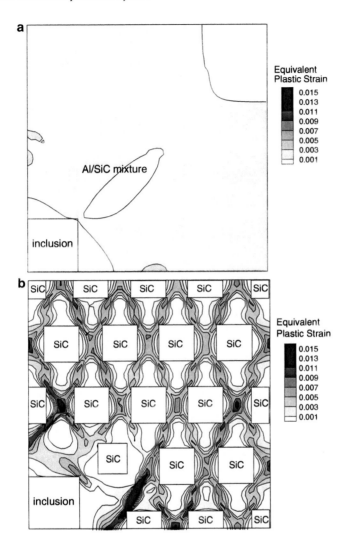

Fig. 6.26 Contour plots of equivalent plastic strain in the composite at an overall applied plastic strain of 0.002, obtained from the (**a**) homogenized and (**b**) explicit models

metal-ceramic composite materials for our discussion. A case study on the rate-dependent viscoelastic matrix composites is given in Sect. 6.3.4.

6.3.1 Basic Consideration

Some analytical models for predicting the composite CTE are first briefly reviewed. The constituent phases are assumed to be isotropic, linearly elastic. For macroscopically isotropic composites, the effective CTE can be uniquely derived from [69–71]

$$\alpha = \alpha_2 + \left(\alpha_1 - \alpha_2\right)\frac{\left(1/K\right) - \left(1/K_2\right)}{\left(1/K_1\right) - \left(1/K_2\right)}, \tag{6.14}$$

where α and K are the CTE and bulk modulus, respectively, and the subscripts 1 and 2 refer to the two phases. The variables without subscript stand for the effective properties of the composite. To obtain a close-form expression of composite CTE, the effective bulk modulus K needs to be known. The Hashin-Shtrikman bounds of the bulk modulus [38] described in Sect. 6.1.3 can serve the purpose. They are written here as

$$K^{upper} = K_2 + f_1 \left[\frac{1}{K_1 - K_2} + \frac{3f_2}{3K_2 + 4G_2}\right]^{-1} \tag{6.15}$$

$$K^{lower} = K_1 + f_2 \left[\frac{1}{K_2 - K_1} + \frac{3f_1}{3K_1 + 4G_1}\right]^{-1}, \tag{6.16}$$

where G is the shear modulus and f is the phase volume fraction. Equations (6.15) and (6.16) hold when $K_1 < K_2$ and $G_1 < G_2$. The combination of (6.14) and (6.16) is also equivalent to the Kerner model [72]. It should be noted that the upper and lower bounds of K result in, respectively, the lower and upper bounds of α.

For a composite material, the effective Young's modulus has been shown to be influenced by the internal phase geometry. This should not be the case for the effective CTE of the composite, however, as (6.14) suggests. The reason is that α depends directly on bulk modulus K, and K is essentially insensitive to the shape and dispersion of the reinforcement [73]. We can use simple numerical modeling on the particle/matrix system to test this notion. The hexagonal packing of circular reinforcement shown in Fig. 6.4 is considered. Note that to accommodate more than one particle in a cell, we use the 2D plane-strain approach, and no mechanical loading is applied. In addition, the three arrangements of square particles shown in Fig. 6.15, namely models A, B and C, are also used. In all models the bottom and left boundaries of the cell quadrant are only allowed to move in the 1- and 2-directions, respectively, and the right and top boundaries are constrained to remain vertical and horizontal, respectively. A temperature increase from 20 to 25°C is imposed, and the overall linear strain is calculated by dividing the resulting dimensional change with the original length. The composite CTE is then simply the overall linear strain per unit temperature difference. With the four models the shape and spatial distribution effects can both be addressed.

The composites are again taken to be an Al matrix reinforced with SiC particles. However, in the current set of thermal expansion modeling (Sects. 6.3.1–6.3.3) the matrix is assumed to be pure Al in an annealed state, and most of the thermo-mechanical properties are temperature-dependent as listed in Table 6.1. Plastic deformation of Al follows linear hardening up to the ultimate strength σ_u, beyond which perfect plasticity ensues. In all models the area fraction of SiC particles is 30%. In the presentation of results the effective CTE values are normalized with the room-temperature CTE of the pure Al matrix under the same plane strain condition, $31.14 \times 10^{-6} K^{-1}$.

Table 6.1 The temperature-dependent thermo-mechanical properties used for the finite element modeling of thermal expansion response of Al/SiC composites

	T (°C)	E (GPa)	ν	α (K^{-1})	σ_y (MPa)	H (MPa)	σ_u (MPa)
Al	20	69.0	0.33	23.0×10^{-6}	34.0	150.0	79.0
	100	–	–	–	32.0	129.3	70.8
	149	–	–	–	29.0	116.7	64.0
	204	–	–	–	24.0	102.3	54.7
	260	–	–	–	18.0	88.0	44.4
	316	–	–	–	14.0	40.0	26.0
	371	–	–	–	11.0	26.0	18.8
	600	47.7	0.33	32.6×10^{-6}	8.0	0	8.0
SiC	20	450.0	0.17	4.7×10^{-6}	–	–	–
	600	437.9	0.17	5.3×10^{-6}	–	–	–

The symbols E, ν, α, σ_y and H stand for Young's modulus, Poisson's ratio, CTE, initial yield strength and linear hardening slope, respectively. The symbol σ_u represents the maximum stress beyond which the material shows perfect plasticity. A linear variation of properties with temperature between the indicated temperatures is assumed

Table 6.2 Effective coefficients of thermal expansion (CTE) obtained from the finite element analysis of various Al/SiC composite models, all based on the 2D plane-strain unit-cell methodology

	Normalized CTE (w/o residual stresses)	Normalized CTE (w/ residual stresses)
Hexagonal (// 1-axis)	0.716	0.708
Hexagonal (// 2-axis)	0.716	0.704
Square, model A	0.712	0.702
Square, model B	0.710	0.702
Square, model C	0.708	0.702

The hexagonal model is the same as in Fig. 6.4, with the CTE results obtained from the dimensional changes along the 1- and 2-axes listed separately. The square models (A, B and C) are identical to those in Fig. 6.15 (all square models have identical results along the 1- and 2-directions). The volume (area) fraction of SiC particles is fixed at 30% in all models. The results are obtained by imposing a temperature increase from 20 to 25°C. The case with the thermal residual stresses is simulated by including a first cooling step from 600 to 20°C. All composite CTE values presented are normalized with the room-temperature CTE of the pure Al matrix under the same plane strain condition ($31.14 \times 10^{-6}\,\mathrm{K}^{-1}$)

The effective normalized CTE values obtained from the finite element modeling are listed in Table 6.2. The center column is the "clean" results from 20 to 25°C, with 20°C taken to be the stress-free initial state. This temperature change is small enough for the soft Al to remain elastic. For the hexagonal model, the CTEs along the two different directions are reported, and they turn out to be equivalent under this elastic state. For the three square models, their response in the 1- and 2-directions are equivalent by default. It can be seen that there are only slight differences in effective CTE among the various models. The right column in Table 6.2 shows the results with prior thermal residual stresses in the composites. This is simulated by first including a cooling process from 600 to 20°C, with 600°C being the stress-free temperature. The determination of composite CTE is from subsequent heating between 20 and 25°C. In all cases the composite CTE is seen to be reduced due to the presence of residual stresses (and prior plastic deformation). Nevertheless, the CTE values among the various models remain even closer. It can thus be concluded that the shape and spatial distribution of the reinforcement plays a very small role in affecting the composite thermal expansion response.

6.3.2 Effect of Phase Contiguity

Although the effective thermal expansion of metal/ceramic composites is basically not influenced by the particle geometry, it is strongly affected by the phase contiguity. Depending on the ceramic content and processing methods, a macroscopically isotropic metal/ceramic composite may be metal-matrix (with isolated ceramic particles as considered in previous sections), ceramic matrix (with isolated metallic particles) or interpenetrating (with both the metallic and ceramic phase being continuous in a 3D fashion). For example, the powder consolidation technique may be used to produce all forms of composites, and the molten metal infiltration technique may be used to produce metal-matrix and interpenetrating composites. The relative phase contents and phase contiguity may be adjusted to tailor the overall CTE of the composite for thermal management applications such as electronic packaging [74–76]. In this section we present a systematic 3D finite element analysis, with special attention devoted to phase contiguity [77].

6.3.2.1 Model Setup

A unified model setup, featuring the three types of phase connectivity extending periodically into the 3D space, is chosen. Figure 6.27a shows the unit cell for the case of a discrete phase (shaded cube) embedded within a continuous phase for modeling metal-matrix and ceramic-matrix composites. Figure 6.27b shows the unit cell for interpenetrating composites where both phases (one of which bounded

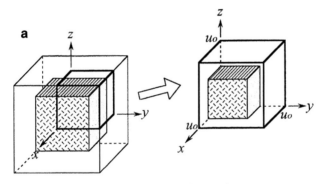

metal-matrix or ceramic-matrix

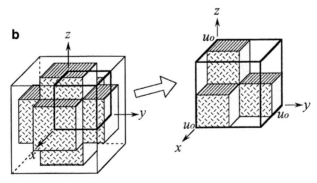

interpenetrating phases

Fig. 6.27 Schematics showing the phase arrangements in unit cells and the cell octants used for the 3D modeling: (**a**) the metal– or ceramic–matrix composites; (**b**) the interpenetrating composite [77]

by the shaded surfaces), having the same geometry, form a three-dimensional network in space. Only an octant of the unit cell, shown at the right-hand side of (a) and (b), needs to be considered for the calculation because of symmetry. The boundary conditions ensure the preservation of mirror symmetry of the periodic structure. As shown in Fig. 6.27, the cell octant analyzed numerically is within the volume $0 \leq x \leq u_0$, $0 \leq y \leq u_0$ and $0 \leq z \leq u_0$. Along the planes $x=0$, $y=0$ and $z=0$, the displacement in the x-, y- and z-directions, respectively, vanishes during deformation caused by the imposed temperature change. The boundary planes $x=u_0$, $y=u_0$ and $z=u_0$ are free to move but have to remain planar and parallel to their original positions for preserving the compatibility with adjacent cells. During deformation $x=u_0$, $y=u_0$ and $z=u_0$ become, respectively, $x=u$, $y=u$ and $z=u$ where u is determined by the analysis. The composite CTE is then defined as $(1/\Delta T) \cdot (u - u_0)/u_0$, where ΔT is the imposed temperature change.

The geometric model adopted here appears to be highly idealized. However, its validity can be justified in light of the fact that the shape and spatial distribution of the discrete phase have essentially no influence on the overall CTE of the composite (Sect. 6.3.1). Therefore, the use of a cubic particle shape in a regular cubic array for the metal-matrix and ceramic-matrix composites (Fig. 6.27a), and the orthotropic type of network for the interpenetrating composite (Fig. 6.27b), should be appropriate for the present purpose. An important advantage of this approach is that the results for different forms of phase connectivity, with the phase fraction from 0 to 100%, can be uniquely simulated and compared in a simple and unambiguous fashion. The thermo-mechanical properties of Al and SiC used in the modeling are listed in Table 6.1. The composite CTE values presented in this section are based on a temperature change from 20 to 100°C, either with an initially stress-free state at 20°C or with a prior processing step of cooling from 600 to 20°C to create built-in thermal residual stresses.

6.3.2.2   Composite CTE Without Thermal Residual Stresses

Results of the composite CTE from simulations without the processing induced thermal residual stresses are first presented. Figure 6.28 shows the composite CTE as a function of SiC volume fraction for the three types of phase connectivity. Distinct differences emerge for the entire range of phase concentration. For instance, at the SiC volume fraction of 0.4, the difference in CTE due to different phase connectivity can be as high as $3.5 \times 10^{-6} \, \text{K}^{-1}$. The metal-matrix and ceramic-matrix cases show, respectively, the greatest and smallest CTE values. This is because Al has a higher CTE and is thus more compliant with respect to the temperature change. When the more compliant phase is the discrete phase, deformation is severely constrained by its less compliant surrounding. Such constraint is much reduced when the more compliant phase forms a continuous matrix. The interpenetrating model is intermediate between the metal-matrix and ceramic-matrix cases, as can be expected. The CTE values of the interpenetrating composite, however, are closer to those of the ceramic-matrix composite. This is due to the fact that in

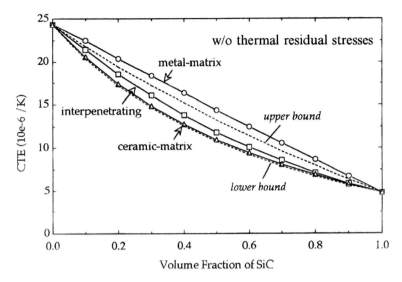

Fig. 6.28 Modeled composite CTE as a function of SiC volume fraction for the case without the presence of thermal residual stresses. Bounds obtained from analytical expressions ((6.14)–(6.16)) are also included for comparison

interpenetrating composites, both phases are continuous in all three dimensions, so the compliant phase is still strongly constrained.

Also included in Fig. 6.28 are the analytical bounds of the composite CTE. The upper bound is the combination of (6.14) and (6.16), and the lower bound the combination of (6.14) and (6.15), with phases 1 and 2 in these equations being Al and SiC, respectively. The *average* thermoelastic properties between 20 and 100°C are used for obtaining the analytical solutions. It is seen that the modeled CTE values of metal-matrix composites are higher than those specified by the upper bound. This mainly stems from the fact that the soft pure Al matrix has actually undergone plastic yielding during the heating process from 20 to 100°C, which cannot be captured by the analytical solutions. Plastic yielding of the continuous metallic phase results in a higher overall CTE value than that in the elastic case [78], because a plastically deforming metal tends to flow, without increasing the stress, to accommodate the thermal expansion mismatch with the ceramic. The contribution of composite CTE from the metal matrix is therefore more easily revealed. It can be seen from Fig. 6.28 that the numerical prediction for the metal-matrix case is quite close to the linear rule-of-mixtures approximation.

In the case of ceramic-matrix composites, the numerical curve in Fig. 6.28 lies just slightly above the analytical lower bound. Since the embedded metal particles are totally confined in this case, essentially the lowest possible composite CTE is to be expected. Furthermore, the metallic phase is severely constrained so that a rather uniform hydrostatic stress state with very little plasticity exists (except near the sharp corners of particles) [73, 78]. As a consequence, the numerical prediction closely matches the lower-bound solution which is based on the *elastic* analysis.

6.3.2.3   Composite CTE with Thermal Residual Stresses

We now consider numerical results on the composite CTE between 20 and 100°C, with a prior processing step from 600 to 20°C included. Figure 6.29 shows the modeled composite CTE as a function SiC volume fraction for the composites with different types of phase connectivity. Both the cases with (solid curves) and without (dashed curves) the thermal residual stresses are included in the figure for comparison. For clarity purpose only the range of volume fraction between 0.2 and 0.8 is shown. It is observed that the incorporation of residual stresses in the modeling significantly alters the composite CTE of all types of phase connectivity for all phase fractions. With the residual stresses accounted for, the metal-matrix composite shows a decreasing trend but the ceramic-matrix and interpenetrating composites show an increasing trend of composite CTE, compared to the case without the residual stresses. The net effect is that the curves for the three types of phase connectivity become closer together, with the largest difference of CTE between the metal-matrix and ceramic-matrix cases being about $1.7 \times 10^{-6}\,\mathrm{K}^{-1}$ (for a SiC volume fraction of 0.5), which is still quite significant.

The opposite trends of CTE modification caused by thermal residual stresses observed in Fig. 6.29 for the metal-matrix composite and for the ceramic-matrix and interpenetrating composites deserve more attention. This is due to the plastic deformation of metal affected by the constraint imposed by the stiff ceramic phase. As discussed above, a greater extent of metal plasticity *during* thermal loading results in a higher CTE of the composite. Therefore we can compute and compare the equivalent plastic strains in Al, occurring *during* heating from 20 to 100°C, for the various forms of Al/SiC composites with and without the thermal residual

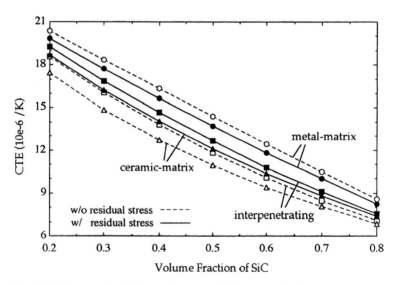

Fig. 6.29 Modeled composite CTE as a function of SiC volume fraction, with and without the thermal residual stresses accounted for

Table 6.3 The volume-averaged accumulated equivalent plastic strains in Al obtained from the finite element modeling, for the three forms of composites with a 50% SiC volume fraction

	Metal-matrix	Ceramic-matrix	Interpenetrating
W/o residual stresses			
20°C→100°C	1.78×10^{-3}	4.96×10^{-6}	4.28×10^{-5}
W/ residual stresses			
600°C→20°C	2.49×10^{-2}	2.18×10^{-3}	8.90×10^{-3}
600°C→20°C→100°C	2.59×10^{-2}	2.20×10^{-3}	9.08×10^{-3}
20°C→100°C	1.0×10^{-3}	2.0×10^{-5}	1.8×10^{-4}

This table is used for comparing the average plastic strains in Al experienced during heating from 20 to 100°C, with or without the thermal residual stresses accounted for

stresses. In the composite the deformation field is highly non-uniform, with plasticity more concentrated near the sharp corners and edges of the interface. Taking the SiC volume fraction of 0.5 as an example, we consider in the following the accumulated equivalent plastic strain *averaged* over the volume of Al, as listed in Table 6.3. In the case without thermal residual stresses (i.e., an initial state at 20°C), a temperature increase of $\Delta T = 80°C$ is already large enough to trigger detectable plastic yielding in Al, although for the ceramic-matrix and interpenetrating composites the plastic strain values are extremely small. In the case with thermal residual stresses (i.e., an initial state at 600°C), strong plasticity exists upon cooling to 20°C, especially in the metal-matrix composite. The constituent phases undergo unloading during the subsequent heating process. At 100°C reversed yielding already commences in Al for all three types of phase connectivity, because larger magnitudes of accumulated plastic strain are seen, in Table 6.3, for the thermal history of 600°C→20°C→100°C, compared to those for 600°C→20°C. The net increases of plastic strain from 20 to 100°C are listed in the bottom row of the table, which shows the extent of plasticity occurring during the heating period used for calculating the composite CTE for the case with thermal residual stresses. A direct comparison of these plastic strain values with those free of thermal residual stresses can now be made. It can be seen from Table 6.3 that upon heating from 20 to 100°C, the gain of the average plastic strain in Al for the metal-matrix composite is higher in the case without the residual stresses (1.78×10^{-3}) than with the residual stresses (1.0×10^{-3}). This is reflected in Fig. 6.29 that a decreasing trend is observed after the existence of residual stresses is included, because stronger metal plasticity generally leads to higher composite CTE values. For the other two forms of phase connectivity, the gains in Al plastic strain in the case without the residual stresses (4.96×10^{-6} and 4.28×10^{-5} for ceramic-matrix and interpenetrating composites, respectively) are smaller than those with the residual stresses included (2.0×10^{-5} and 1.8×10^{-4} for ceramic-matrix and interpenetrating composites, respectively). Therefore, in Fig. 6.29, an increasing trend is observed for these two types of phase connectivity after the residual stresses are included.

The analysis above illustrates the correlation between the plastic behavior of the metallic phase and the overall thermal expansion response of metal-ceramic composites. Plastic deformation in the metal is strongly influenced by the constraint induced by the different phase arrangements. The fact that the metal-matrix composite shows an opposite trend compared to the other two composites can be attributed, at least qualitatively, to the much higher plastic strain in the metal matrix after the initial cool-down from 600°C (Table 6.3). Since the higher plastic strain during forward loading renders a greater magnitude of yield stress for the reversed loading (due to the isotropic hardening model used in the modeling), the total amount of reversed yielding in the metal matrix is likely to be reduced to a large extent. This is consistent with what was observed in Table 6.3. The modeling also illustrates, in a realistic three-dimensional manner, that the metallic phase in interpenetrating composites experiences a similar type of triaxial constraint as in ceramic-matrix composites. The CTE values of interpenetrating composites are closer to those of ceramic-matrix composites, and more importantly, the same trend of CTE change caused by metal plasticity are observed for these two classes of composites during reversed thermal loading. Therefore, it can be deduced that, treating the interpenetrating composite as a metallic phase embedded within a continuous ceramic matrix using 2D modeling, for purposes of computational efficiency and ease of visualization, is a feasible approach for analyzing the qualitative thermal loading behavior. In the following section, this type of 2D analysis taking into account small voids in high-ceramic-content composites is presented.

6.3.3 Composites with Imperfect Metal Filling

In actual metal-ceramic composites with high ceramic contents, small voids frequently exist at the contact points between ceramic particles or sharp concave corners of the continuous ceramic phase [79]. This is mainly due to the imperfect filling of metal during fabrication, and can have a direct impact on the overall thermal expansion response of the composite. The 2D unit-cell model may again be used to address the problem [79, 80].

6.3.3.1 Model Setup

A representative configuration featuring a periodic arrangement of discrete Al phase embedded within continuous SiC is shown in Fig. 6.30. This model enables easy incorporation of voids as well as SiC particle percolation. With the presence of cracks, the model actually represents high-particle-concentration metal-matrix composites with contacts between ceramic particles. When there is no crack the model then simulates ceramic-matrix or interpenetrating composites. The generalized plane strain formulation is used in the present set of modeling. Predictions based on the present 2D calculations cannot be directly compared with the experimentally measured CTE values; however, they do allow substantiation of qualitative trends and the effect of microscopic deformation mechanisms as presented below. As before attention is confined to deformations that preserve the mirror

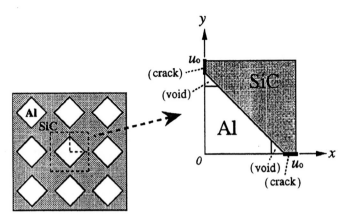

Fig. 6.30 Overall arrangement of the Al and SiC phases and the quadrant of a unit cell used for the modeling. The possible inclusion of cracks and/or voids in the model is indicated [79]

symmetry of the structure. Along the axis $x=0$, the displacement in the x-direction vanishes. Along the axis $y=0$, the displacement in the y-direction vanishes. The external horizontal and vertical boundaries of the cell quadrant are free to move but have to remain, respectively, horizontal and vertical. The thermo-mechanical properties of Al and SiC used in the model are listed in Table 6.1. All the calculations start with an initial cooling to 20°C from a stress-free temperature of 600°C, with subsequent cycling between 20 and 320°C.

We first present a baseline result where no void or crack is included in the model. The area fractions of SiC and Al are taken to be 0.55 and 0.45, respectively. Figure 6.31 shows the simulated response (marked "elastic-plastic Al") of the overall thermal strain of the composite as a function of temperature. It is seen that thermal cycling causes essentially no hysteresis. This is because plastic deformation in Al is strongly constrained by the surroundings. Yielding occurs only in regions very close to the sharp corners. Also shown in Fig. 6.31 is the result when the Al phase is taken to be purely elastic (marked "purely elastic Al" in the figure). Note that the response is not linear because of the variation of material properties with temperature. It is clear by comparing these two curves that the instantaneous composite CTE (slope of the strain–temperature curve) shows a higher value when metal plasticity is included in the model. This observation is consistent with the 3D modeling result in Sect. 6.3.2.

6.3.3.2 Effect of Voids

We now consider the void-containing composite where the area fractions at the stress-free state of SiC, Al and void are 0.55, 0.42 and 0.03, respectively. The SiC phase is continuous because there is no crack included. Figure 6.32 shows the modeled variation of strain with temperature. Because of the very low yield strength of

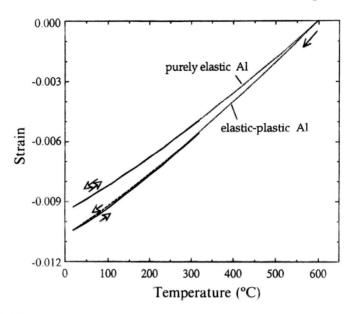

Fig. 6.31 Modeled thermal expansion (strain) curves for the thermal history of 600°C → 20°C → 320°C → 20°C. No voids or cracks are included in the modeling

Al at high temperatures, apparent yielding of the composite begins at about 585°C during initial cooling. Note, by comparison with Fig. 6.31, that the overall strain after initial cooling is substantially reduced in the presence of voids. This is because metal around the voids is much less constrained, so that deformation can be accommodated without the need for significantly changing the macroscopic strain. In Fig. 6.32 strong hysteresis is seen. In general, two linear portions exist during heating and cooling between 20 and 320°C. For example, point B labeled in the figure represents the transition of slope at about 150°C for the first heating cycle. It is noted that, after initial cooling to point A, the metal has experienced extensive plastic yielding, with severe deformation near the Al-SiC-void triple junction. The segment AB represents essentially elastic response of the composite. A significant increase in plasticity occurs between B and C especially near the void site (this can be easily confirmed if one examines the contour plots resulting from the modeling). Upon cooling from 320°C, the Al phase undergoes elastic and then plastic deformation, so hysteresis is observed. In Fig. 6.32 the strain–temperature curve for the second cycle between 20 and 320°C is also shown. The transition from elastic to plastic responses during heating is postponed to a higher temperature because of the cyclic hardening of the metal (the isotropic hardening model is used).

In order to check the variation of void configuration at different stages, the deformed finite-element meshes corresponding to points A, B and C in Fig. 6.32 during first heating are shown and overlaid on that of the initial stress-free state (at 600°C) in Fig. 6.33a, b and c, respectively. After initial cooling to room

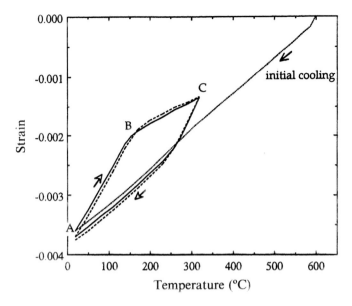

Fig. 6.32 Modeled thermal expansion (strain) curves for the void-containing, crack-free composite. The thermal history involved is 600°C → 20°C → 320°C → 20°C → 320°C → 20°C

temperature, the void size increases, because the metal tends to contract more than the surrounding ceramic (Fig. 6.33a). Upon subsequent heating to B, the elastic deformation does not cause significant void shrinkage (comparing Fig. 6.33b and a). At stage C, however, the metal-void front has moved toward the corner much more deeply, due to the strong plastic flow around the void (comparing Fig. 6.33c and b). Therefore there is no need for the composite to change its external dimensions as much, so a smaller change of strain with temperature from points B to C in Fig. 6.32 is seen.

6.3.3.3 Effect of Particle Contact

The effect of reinforcement contact in the SiC particle-reinforced Al matrix composite (metal-matrix) is now examined. As mentioned above, this is modeled by placing a crack at the junction of two SiC particles (Fig. 6.30). The additional boundary conditions imposed are such that the cracks remain closed (the crack face lies on the symmetry axis in the model) if the crack face is experiencing a compressive normal traction, but the cracks are free to open when the compressive normal traction is reduced to zero. For the SiC and Al area fractions of 0.55 and 0.45, respectively (i.e., no void), predictions for the composite with cracks show exactly the same strain–temperature response as in the case without cracks (the "elastic-plastic Al" case in Fig. 6.31). During initial cooling, very high compressive stresses are generated

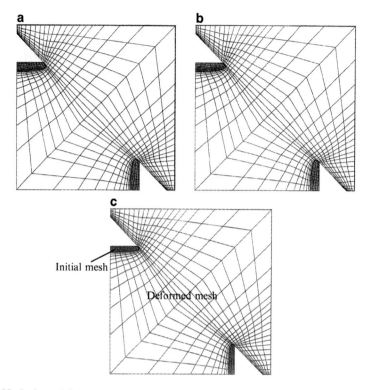

Fig. 6.33 Deformed finite element meshes at stages (**a**) A, (**b**) B and (**c**) C labeled in Fig. 6.32 overlaid on the initial stress-free mesh for the void-containing, crack-free composite

near the crack face, because of the strongly contracting Al near the corner. Upon subsequent heating, the stresses are never relieved completely within the temperature range considered, so the cracks remain closed throughout the cyclic history.

If, in addition to cracks, voids are made existent at the metal corners, the modeled composite thermal expansion becomes that shown in Fig. 6.34, when the void fraction is 0.03 (0.55 and 0.42 for SiC and Al, respectively). After initial cooling, the crack face is under compression. Since Al is now replaced with a void at SiC corners, the compressive stresses are not as large as in the void-free composite. These compressive stresses are then reduced during heating. At around 95°C, the stress at the crack face vanishes so crack faces can be pushed apart (due to the expansion of Al) thereafter, rendering a more compliant composite response. Hence, an increase in CTE (slope of the strain–temperature curve) is seen in Fig. 6.34 during the heating phase. Note that crack opening occurs before voids begin to shrink (as in the crack-free void-containing case), so the decrease in CTE during heating shown in Fig. 6.32 no longer happens here. During cooling from 320°C, the crack faces are pulled closer together by the contracting metal. At around 85°C, crack closure occurs, and a decrease in CTE is obtained upon further cooling because compressive stresses can now be transmitted across the crack.

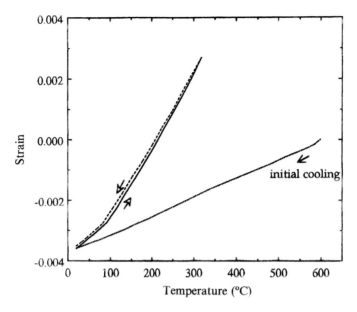

Fig. 6.34 Modeled thermal expansion (strain) curves for the composite with voids and cracks at the particle contact areas. The thermal history is 600°C → 20°C → 320°C → 20°C

6.3.3.4 Comparison with Experiment

The above analyses have revealed the interesting effects caused by the presence of voids and/or cracks in the simple 2D model. Experimental measurements of thermal expansion of Al/SiC composites with high SiC contents have actually shown some of the characteristics in the modeling [79]. For instance, the SiC foam-reinforced (inter-penetrating) composite, corresponding to the modeling case with voids but no cracks, displayed a bi-linear type of strain–temperature response during heating and cooling, with hysteresis behavior observed for a full cycle. This is qualitatively consistent with the modeling result in Fig. 6.32. For the SiC particle-reinforced composite (corresponding to the modeling case with voids and cracks), a bi-linear type of strain–temperature response was observed during heating and cooling, with a higher CTE value at elevated temperatures. This is also in qualitative agreement with the modeling result presented in Fig. 6.34. The present section again demonstrates the usefulness of *simple* numerical modeling in helping to rationalize actual material response or experimental findings which may not be readily explained otherwise. The key is to design an adaptable model which is easily implemented and yet physically meaningful.

6.3.4 Viscoelastic Matrix Composites

This section is also devoted to the effective thermal expansion of particle-filled composites, but with the time-dependent viscoelastic matrix material. One technologically important example of this type of composites is the epoxy resin containing high-fraction silica (SiO_2) particle fillers, used as transfer molding compounds to encapsulate microelectronic devices [81, 82]. One of the main purposes of incorporating fillers is to reduce the overall CTE of the encapsulant, and hence reduce the thermal expansion mismatch between the encapsulant and the encapsulated parts. (An unfilled epoxy has a typical CTE value of $50 \times 10^{-6} K^{-1}$.) The filler may take the form of solid particles or hollow spheres (the so called micro-balloons). In this section, we present numerical characterization of the CTE of SiO_2 filled epoxy. In particular, to provide a different perspective from the 3D and 2D planar models used in Sects. 6.3.1–6.3.3, the axisymmetric unit-cell approach treated extensively in mechanical loading is now re-employed.

6.3.4.1 Model Setup

Figure 6.35a shows a schematic of the axisymmetric unit cell and the quadrant used in the modeling. The filler (particle) is assumed to be either a solid sphere or a hollow sphere. In the case of a hollow sphere, the calculation of filler volume fraction includes the solid shell portion and the embedded void. Two forms of filler arrangement are considered. The first arrangement consists of a periodic array of particles, as in Fig. 6.35b where the particle layout on a representative rz-plane is shown. The corresponding cell quadrant is highlighted in the figure. The boundary conditions are such that, along the r- and z-axes, displacements are allowed only in the r- and z-directions, respectively, and the top and right boundaries can move only in a self-parallel manner for preserving compatibility with adjacent cells. As a consequence,

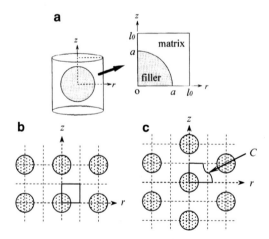

Fig. 6.35 (a) The axisymmetric unit cell and the quadrant used in the modeling. The *shaded region* refers to the filler (particle), being either a solid sphere or a hollow sphere. (b) and (c) Schematic views of filler arrangements on a representative rz-plane: (b) the aligned arrangement; (c) the staggered arrangement [83]

during deformation $r = l_0$ becomes $r = l_r$ and $z = l_0$ becomes $z = l_z$, where l_r and l_z are determined by the analysis. It should be noted that the axisymmetric model does not generally lead to an isotropic thermal response so $l_r \neq l_z$. The modeled composite CTE presented below will be the average of the values along the axial and two lateral directions, defined to be $\{2[l_r - l_0]/l_0 + [l_z - l_0]/l_0\}/3\Delta T$, where ΔT is the temperature change imposed [83].

The second arrangement of filler particles takes into account the staggered distribution depicted in Fig. 6.35c. The cell quadrant for the axisymmetric modeling appears the same, but the boundary conditions are different. The bottom, left and top boundaries follow the same boundary conditions as in the first arrangement. The right boundary ($r = l_0$ initially) is constrained by the consideration of compatibility with the neighboring cell, which is related to the original cell through an inversion center C in the middle of the boundary as shown in Fig. 6.35c. Therefore, on the right boundary, the displacement along the z-direction, (u_z) is constrained by

$$u_z(p) + u_z(q) = 2u_z(C) \tag{6.17}$$

where p and q denote pairs of points on $r = l_0$ which are equidistant from the top and bottom of the cell quadrant, respectively. The center point C itself also constitutes such a pair. The displacement along the r-direction (u_r) is constrained by

$$[l_0 + u_r(p)]^2 + [l_0 + u_r(q)]^2 = 2[l_0 + u_r(C)]^2. \tag{6.18}$$

Equation (6.17) is required to bring the edge displacements along the axial direction between two neighboring cells in agreement, and (6.18) assures compatibility in the radial direction such that the total cross-sectional area is independent of the z-coordinate [41]. The composite CTE is then defined as $\{2u_r(C)/l_0 + [l_z - l_0]/l_0\}/3\Delta T$.

In the model the silica phase is assumed to be elastic, with Young's modulus 72 GPa, Poisson's ratio 0.16 and CTE $0.5 \times 10^{-6} \, \mathrm{K}^{-1}$. The epoxy phase is modeled as a simple linear viscoelastic solid, known as the three-element standard linear solid model shown schematically in Fig. 6.36 (see also Sect. 2.4). The parameters k_1 ($=0.1$ GPa) and k_2 ($=2.0$ GPa) are the spring constants and η ($=20.0$ GPa s) is the dashpot viscosity. This gives an instantaneous Young's modulus of 2.1 GPa which is typical of epoxy materials. Within the context of linear viscoelasticity, the time-dependent response can be described as the relaxation behavior of the elastic modulus,

$$M = M_u - \{M_u - M_r\}\{1 - \exp[-t/\tau_M]\} \tag{6.19}$$

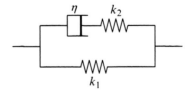

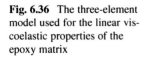

Fig. 6.36 The three-element model used for the linear viscoelastic properties of the epoxy matrix

where M_u and M_r stand for the unrelaxed (instantaneous) and relaxed (long-term) moduli, respectively, t is time, and τ_M is the characteristic relaxation time for the modulus M. Equation (6.19) can be applied to both the shear and bulk moduli, although in the present analysis the bulk modulus is assumed to be time-independent ($= 34.6$ GPa) so no volumetric relaxation occurs. Based on the spring and dashpot parameters chosen, the unrelaxed and relaxed shear moduli are then 0.7047 and 0.0333 GPa, respectively, and the characteristic relaxation time for the shear modulus is 9.936 s. The viscoelastic response of the matrix is thus completely defined. This linear viscoelastic model can be viewed as "long-term elastic" in that, after having been subjected to a constant strain for a long time, the response settles down to a constant stress. The CTE value of epoxy is taken to be $50.0 \times 10^{-6} \, \text{K}^{-1}$. All material parameters used in this section are independent of temperature. In the modeling thermal expansion is obtained by imposing a spatially uniform temperature increase of $\Delta T = 80°C$, with constant heating rates of 10°C/s or 1°C/s. The composite CTE defined above is thus the derived quantity over a temperature span of 80°C.

6.3.4.2 Composites with Solid Fillers

We first present an assessment of the model prediction of CTE using the axisymmetric approach, by comparing the modeling results with analytical solutions. Here only the case of time-independent elastic response (with the matrix moduli being those of the instantaneous values) and solid sphere filler is considered. The analytical expressions follow (6.14)–(6.16) described in Sect. 6.3.1. The lower and upper bounds of the composite bulk modulus K result in the upper and lower bounds, respectively, of composite CTE α. In the present case, however, the material properties chosen are such that the analytical upper and lower bounds practically coincide, because the bulk moduli for epoxy (34.6 GPa) and silica (35.3 GPa) are almost identical. This should put the numerical model to a stringent test because the calculated CTE at a given filler fraction will be compared with a single value instead of a possibly wide range defined by the bounds. Figure 6.37 shows such a comparison. The bounds appear in the figure as a single curve labeled "analytical." The maximum filler volume fraction considered is 60% since this is approximately the highest possible fraction for the sphere-in-cylinder model. It can be seen, as expected, that the CTE value decreases with increasing filler concentration. The numerical results are quite close to the analytical prediction, especially for the case of aligned filler particles. The validity of the present numerical approach is thus demonstrated.

Figure 6.38 shows the modeled composite CTE as a function of filler volume fraction under the heating rates of 10°C/s and 1°C/s. Both the aligned and staggered arrangements of solid spheres are considered. For comparison purpose, calculations assuming a time-independent elastic matrix are also included in the figure. It can be seen that, within the range of heating rates considered, the viscous behavior of epoxy has essentially no effects on the overall CTE, because the CTE differences

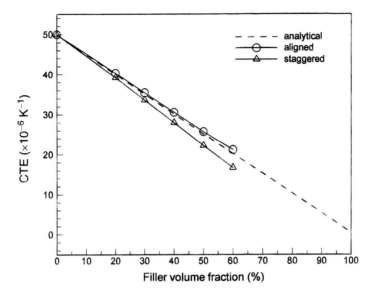

Fig. 6.37 Modeled composite CTE (labeled as "aligned" and "staggered") and the analytical predictions (labeled as "analytical," which actually include the essentially coincidental upper and lower bounds). Only the elastic case in the numerical model is considered

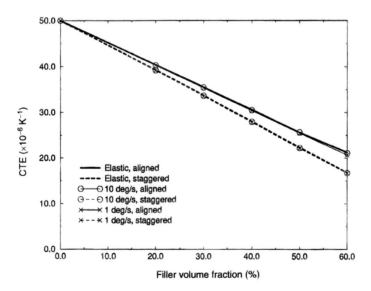

Fig. 6.38 Modeled composite CTE as a function of the volume fraction of solid filler particles. Both the aligned and staggered arrangements are considered

resulting from different heating rates and from the elastic and viscoelastic cases are negligibly small. The composite CTE values for the aligned fillers are greater than those for the staggered fillers. This difference due to particle arrangement is, however, moderate, judged from the fact that there is an extremely large difference in CTE of the matrix phase and the filler phase. This means that the effects of spatial distribution of particles on the composite CTE are small, similar to the case of particle-reinforced metal matrix composites discussed in Sect. 6.3.1.

6.3.4.3   Composites with Hollow-Sphere Fillers

Attention is now shifted to the case with hollow-sphere fillers. Figure 6.39 shows the modeled composite CTE as a function of filler volume fraction, for hollow particles of thickness ratio of 0.1. (The thickness ratio is defined as the ratio of the wall thickness and radius of the particle.) Again heating rates of 10 and 1°C/s as well as the instantaneous elastic case are included. It can be seen that aligned particles result in moderately higher CTE values than staggered particles, as in the case of solid spheres. The effect of heating rate is small, with a higher heating rate leading to slightly higher CTE values. Treating the matrix as a linear elastic phase gives very small errors of composite CTE, within the range of heating rates considered here.

Figure 6.40 shows the modeled composite CTE as a function of thickness ratio of hollow filler, for a fixed filler volume fraction of 50%. The thickness ratio of unity corresponds to the case of solid particles. Pure voids in the epoxy matrix (thickness ratio of 0) lead to the same CTE of the matrix. It can be observed from Fig. 6.40 that

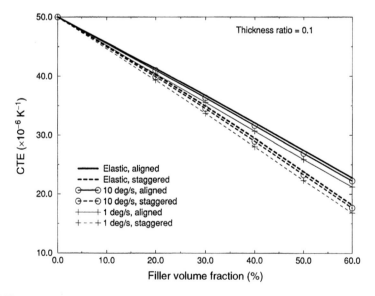

Fig. 6.39 Modeled composite CTE as a function of the volume fraction of hollow-sphere filler particles. Both the aligned and staggered arrangements are considered. The filler thickness ratio is 0.1

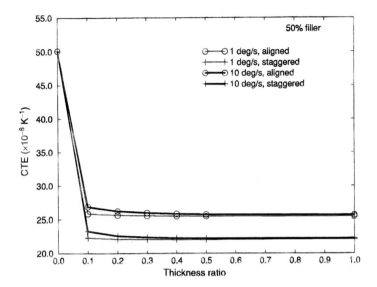

Fig. 6.40 Modeled composite CTE as a function of thickness ratio of the hollow-sphere particles. Both the aligned and staggered arrangements are considered

when the thickness ratio is greater than about 0.1, the composite CTE values become very close to those of the solid-filler composites. The aligned particles result in moderately higher CTE values than the staggered particles in all cases. The effect of heating rate is very small; only at low thickness ratios the 10°C/s heating rate gives rise to slightly higher composite CTE values than the 1°C/s rate.

From the results presented above, it appears that the heating rate does not seem to play an important role in affecting the composite CTE. However, we point out that under the same modeling conditions, the evolution of internal stresses are indeed a strong function of the heating rate. A higher heating rate results in greater stress magnitudes in the epoxy matrix, which will have implications to the initiation of damage in the material. During isothermal hold, significant stress relaxation also occurs. Although the stress contour plots are not presented here, interested readers are encouraged to undertake the modeling analysis for further exploration.

6.4   Internal Deformation Pattern and Damage Implication

During loading of a two-phase composite structure, the macroscopic material response is dictated by the internal deformation features. In Sect. 6.2 we have examined in detail the evolution of local deformation fields in particle-matrix composite systems. In the presentation below we focus on the deformation field and its implications to damage initiation inside the material. Different types of phase structure and/or loading mode from those considered in the previous sections are also explored. The discussion uses case studies concerning mechanical reliability of solder alloys in electronic packaging. In Sect. 6.4.1 the "coarseness" of a lamellar microstructure and its effects on cyclic deformation behavior are studied. In Sect. 6.4.2 the possible improvement in ductility of solder joint due to dispersed particles is investigated.

6.4.1 Cyclic Deformation in Fine and Coarse Lamellar Structure

Tin (Sn)-lead (Pb) alloys with near eutectic compositions have been widely used as soldering materials in the package of semiconductor and communication devices. One of the main reliability issues of great concern is thermo-mechanical fatigue (see also Chap. 5). Stresses in solders are generated due to the thermal expansion difference of the components they connect. Material degradation is caused not only by cyclic straining, but also by creep damage because of the inherent low melting point such that even room temperature is sufficient to trigger high temperature deformation mechanisms. These are further compounded by the fact that the alloy microstructure is changing over time, even at room temperature, with or without applied stresses. A common type of damage is the formation of a microstructurally coarsened band (with greater sizes of phase domains and grains) along which a crack eventually forms, leading to ultimate failure [84–89]. From a metallurgical viewpoint, a coarser microstructure is typically thought to be weaker in response to mechanical loading, i.e., the Hall–Petch type behavior described in materials science textbooks [90, 91]. If this can be applied to thermo-mechanical fatigue of Sn-Pb solder, it would be natural to argue that once a coarsened band has formed, further deformation should become concentrated along the band due to its low strength, with eventual failure along the band. However, this simple view may not accurately reflect what occurs, because the relation between strength and phase size of eutectic Sn-Pb is still subject to great uncertainty [92].

In the presentation below we adopt the unit-cell approach to numerically examine how the phase morphology itself affects the overall mechanical response and local deformation pattern. Idealized "fine" and "coarse" structures in the similitude of lamellar arrangements typical of eutectic alloys are devised. A mechanistic rationale is provided for the experimental fact that fatigue damage tends to develop within the coarsened region. This approach, albeit simple, is very useful in gaining insight into the failure initiation in solder joints, which is a very complex problem by nature.

6.4.1.1 Model Setup

The typical microstructure of an eutectic Sn-Pb alloy consists of alternating layers of a Pb-rich α-phase solid solution and a Sn-rich β-phase solid solution, with the Pb-rich phase appearing to be embedded within the Sn-rich matrix. It may be

Fig. 6.41 Schematics of the (**a**) fine and (**b**) coarse structures used in the modeling [92]

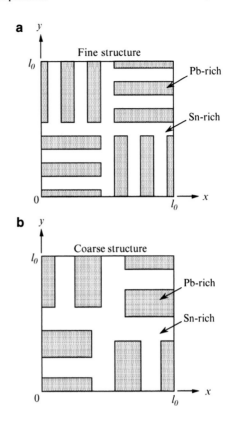

viewed as a composite with ductile metal matrix containing another ductile metal phase. Layers oriented in essentially the same direction form a "colony." In the present model a periodic and regular arrangement of the colony structure is assumed. Figure 6.41a and b shows the model geometry for the "fine" and "coarse" structures, respectively. As before the two-dimensional computational domain is actually a quadrant of a unit cell, and is taken to be a square of side length l_0. Mirror symmetry exists across the four boundaries. Therefore the model consists of four regions each representing a quarter of a colony. In the fine and coarse structures, the aspect ratios of each Pb-rich layer are arbitrarily taken to be 9 and 3.75, respectively. A total of five and three layers of Pb-rich phase exist in each colony for the fine and coarse structures, respectively. An area fraction of Pb-rich phase of 0.45 is assumed in both the fine and coarse structures in this 2D plane-strain model. It is worth emphasizing that the present concern is not to simulate any structure containing a coarsened region surrounded by a finely structured matrix. Rather, separate mechanical characterization of a "coarse" structure and a "fine" structure, each with a spatially extended two-phase layout, is undertaken [92].

During deformation the bottom and left boundaries of the cell quadrant are allowed to move in only the x- and y-directions, respectively. In addition to the uniaxial loading considered extensively in previous sections, the pure shear type of loading is simulated. (In typical solder joint the dominant deformation mode is shear, cf. Chap. 5.) For pure shear the applied normal displacements are imposed on the right and top boundaries, one in compression and the other in tension pre-scribed independently but synchronously. This gives essentially the pure shear condition. For uniaxial loading the prescribed displacement is imposed on the top boundary, and the right boundary is movable but constrained to remain vertical. No thermal effect is considered. Both phases are characterized as elastic-perfectly plastic solids. For simplicity the temperature- and time-dependent plastic behavior for both the actual Sn-rich and Pb-rich phases is not incorporated (and indeed there is no need, for the present purpose). The material properties of the Sn-rich and Pb-rich phases used in the modeling are: $E_{Sn}=50$ GPa, $v_{Sn}=0.33$, $\sigma_{o,Sn}=11$ MPa, $E_{Pb}=28.1$ GPa, $v_{Pb}=0.39$ and $\sigma_{o,Pb}=10$ MPa, where E, v and σ_o stand for Young's modulus, Poisson's ratio and yield strength, respectively.

6.4.1.2 Macroscopic Response and Local Deformation Pattern

Cyclic pure shearing loading is simulated between the nominal shear strains of −0.0025 and 0.0025. The nominal strain is defined to be the prescribed boundary displacement divided by the initial side length of the cell quadrant. Figure 6.42

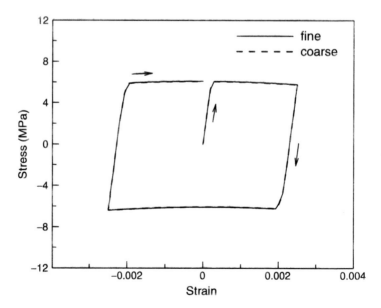

Fig. 6.42 Modeled overall nominal shear stress–shear strain response during the first cycle of pure shear loading

shows the curves of overall nominal shear stress versus shear strain, for both the fine and coarse structures. Only the first loading cycle is shown, since further cycling simply results in repeated hysteresis loops. It is clear that the fine and coarse structures follow essentially the same response. Therefore, one can infer that if a locally coarsened microstructure has developed within the initially fine structure, the coarsened region is neither a weaker nor a stronger part of the material. Of course, this conclusion is predicated upon the validity of the present continuum approach along with the underlying assumptions.

We now turn to the following question: Is the current model still able to rationalize the fact that mechanical damage preferentially occurs within the coarsened region? Figure 6.43a and b show the contour plots of von Mises effective stress and equivalent plastic strain, respectively, for the fine structure at the end of the first loading cycle. The corresponding plots for the coarse structure are shown in Fig. 6.44. In both cases

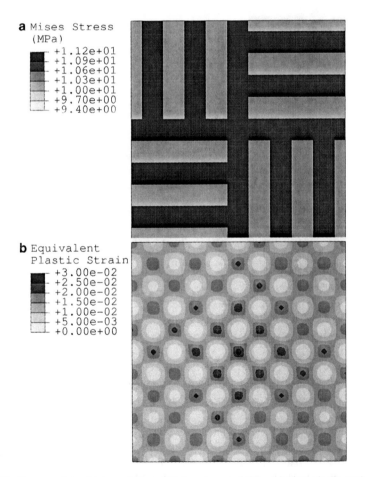

Fig. 6.43 Contour plots of (**a**) von Mises effective stress and (**b**) equivalent plastic strain in the fine structure, at the end of the first pure shear loading cycle

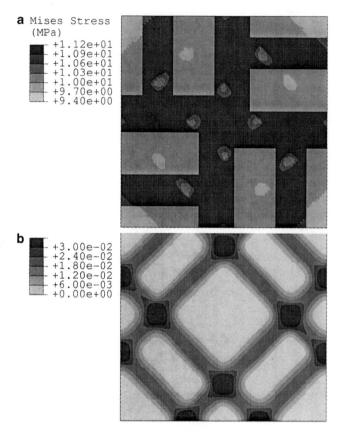

Fig. 6.44 Contour plots of (**a**) von Mises effective stress and (**b**) equivalent plastic strain in the coarse structure, at the end of the first pure shear loading cycle

the entire material has undergone plastic deformation. The local deformation pattern, however, is highly non-uniform. In the fine structure (Fig. 6.43), the von Mises effective stresses are at the levels of yield strengths of the two respective phases. The plastic flow path is severely disturbed so deformation tends to localize into many small regions. In the coarse structure (Fig. 6.44), unloading has actually occurred at some localized spots so the von Mises effective stress contours are not as uniform. Plastic flow path appears as a banded structure. Note the locally unloaded regions in Fig. 6.44a fall inside the lightest contour shade of plastic strain in Fig. 6.44b. In both the fine and coarse structures, the maximum values of equivalent plastic strain are in the Sn-rich phase. In fact, immediately from the onset of plastic yielding, the maximum plastic strain appears at the same location throughout the cyclic deformation history. Therefore one can make a direct comparison of the maximum plastic strain during deformation for the two structures.

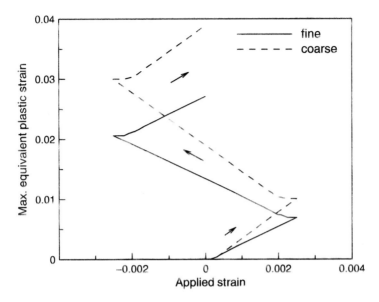

Fig. 6.45 Variations of the local maximum equivalent plastic strain with the applied macroscopic strain during the first cycle of pure shear loading

Figure 6.45 shows the variations of maximum local equivalent plastic strain with the applied macroscopic shear strain, for the fine and coarse structures during the first full cycle. The very short horizontal segment at the beginning of each loading/unloading phase is due to the elastic response. It is evident that the coarse structure shows a higher level of equivalent plastic strain than the fine structure throughout the history. The difference increases as the deformation progresses. This trend continues with further cycling, which is not shown in the figure. Although the fine and coarse structures show essentially the same macroscopic response (Fig. 6.42), the local evolution of plastic strain in the coarse structure is increasingly stronger than that in the fine structure (Fig. 6.45). The equivalent plastic strain, representing the accumulation of irrecoverable deformation in a ductile material, is a parameter directly related to the propensity of damage initiation (e.g., formation of microvoids or microcracks). It can thus be inferred that, if a coarse structure has formed and is subject to the same macroscopic loading condition as in the surrounding fine structure, the coarse structure is more prone to failure. This can qualitatively explain typical experimental observations of thermo-mechanical fatigue in Sn-Pb solder joints.

For uniaxial loading, the tension-compression type of cyclic history is simulated with the same fine and coarse structures. The resulting macroscopic and local deformation characteristics are qualitatively the same as those in Figs. 6.42–6.45. The coarse structure shows significantly stronger plasticity than the fine structure, and the difference increases further with further cycling just as in the case of pure shear.

6.4.1.3 Model Generalization

The analyses above have shown that a coarser microstructure is neither stronger nor weaker than a finer microstructure in terms of the overall stress–strain response. The true difference lies in the maximum equivalent plastic strain at the microscopic level. Although we do not consider how the coarsened band is formed in actual Sn-Pb solder joints, once they are formed the coarsened region is expected to experience greater plasticity leading to easier damage initiation. Since the present modeling concerns the two-phase microstructure within the continuum framework, it actually bears some resemblance to the micromechanical models for metal matrix composites considered in Sects. 6.1 and 6.2, with the current Pb-rich phase viewed as the reinforcing inclusions. The primary difference is that the inclusion phase in metal matrix composites is normally a purely elastic material with high stiffness. In the present case both the Pb-rich and Sn-rich phases are ductile. Furthermore, in unit-cell composites modeling the inclusion layout is normally not as complex. It was illustrated in Sect. 6.2 that, in metal matrix composites, the macroscopic stress–strain response is strongly influenced by the shape and spatial distribution of the inclusion phase. Curiously, there is virtually no difference in the present Sn-Pb case. We may explore this effect by taking the Pb-rich "inclusions" to be purely elastic or elastic-plastic with a much higher yield strength, with all the other modeling parameters remaining the same. Figure 6.46 shows the overall stress–strain curves resulting from such modeling, for both the fine and coarse structures under

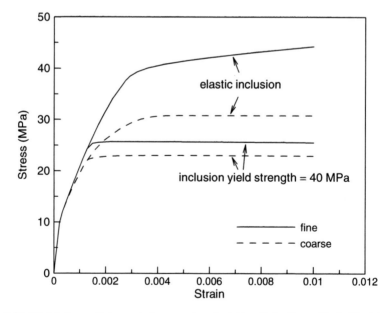

Fig. 6.46 Modeled overall stress–strain curves in uniaxial tensile loading, with the Pb-rich phase taken to be a purely elastic inclusion phase and an elastic-plastic inclusion phase with yield strength 40 MPa

uniaxial tensile loading. For the elastic-plastic inclusion, the yield strength is taken to be 40 MPa (instead of the 10 MPa for the Pb-rich phase). It is clear that, with a purely elastic inclusion phase, the difference in plastic flow stress between the fine and coarse structures becomes very large. For an inclusion phase of yield strength 40 MPa, the difference becomes smaller. Thus, for the previous Sn-Pb model, the same macroscopic response obtained for the fine and coarse structures is due to the very soft nature of the Pb-rich phase, as well as the almost equally soft Sn-rich matrix. Nevertheless, such a small plastic mismatch, along with the somewhat larger elastic mismatch, still results in a significant difference in the evolution of *local* equivalent plastic strain.

As a side note, the two curves in Fig. 6.46 labeled "elastic inclusion" can be used to illustrate a metallurgical feature in another important alloy system. In carbon steels, i.e., an iron (Fe)-carbon (C) alloy, the micro-constituent "pearlite" is an *eutectoid* structure composed of alternating layers of cementite (Fe_3C) and ferrite matrix (Fe-rich solid solution) pertaining to a colony-type arrangement. The phase morphology is essentially the same as the eutectic structure. Within the framework of the present model, if one thinks of the inclusion as cementite (elastic) and the matrix as ferrite (elastic-plastic), then Fig. 6.46 illustrates, from a micromechanical viewpoint, that a finer structure is stronger than a coarser structure. Indeed, by controlling the heat treatment, metallurgists can produce "fine pearlite" which shows a higher strength than "coarse pearlite."

Figure 6.47a and b show the contour plots of equivalent plastic strain in the fine and coarse structures, respectively, with the purely elastic inclusion at the applied tensile strain of 0.01. The plastic flow fields are complicated, with bands in the 45° directions heavily disturbed by the interfaces. The maximum local plastic strain level inside the fine structure is greater than that in the coarse counterpart. Note that this is opposite to the case of eutectic Sn-Pb model. On the basis of plasticity-induced damage initiation in the ductile matrix, the fine structure with elastic inclusions (e.g., "fine pearlite") should become more prone to failure. This is in fact consistent with the experimental trend, showing lower ductility of "fine pearlite" compared to "coarse pearlite" [90].

6.4.2 Particle Dispersion and Ductility Enhancement

In general, reinforcing a ductile matrix with stiff particles results in composite strengthening. However, ductility will be reduced due partly to the localization of plastic flow field within which an elevated plastic strain level commonly exists, leading to easier damage initiation. In this section we illustrate that this is not always the case. Depending on the overall loading configuration, dispersion of particles can sometimes diffuse the initially concentrated deformation field and thus help with delaying damage.

The example is motivated by experimental studies on Sn-based solder alloys with rare-earth element additions. For instance, the addition of lanthanum (La) to

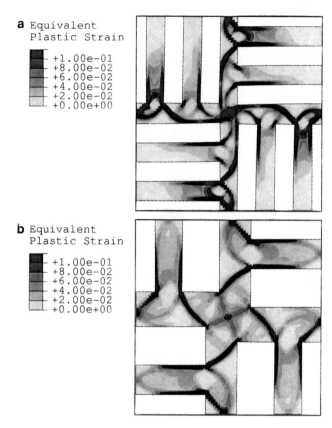

Fig. 6.47 Contour plots of equivalent plastic strain in the (**a**) fine and (**b**) coarse structures when the applied tensile strain is 0.01

form fine LaSn$_3$ particles in Sn-Ag-Cu solder has been shown to increase the strain-to-failure during the lap-shear test [93]. We may now make use of the solder lap-shear model introduced in Chap. 5, to provide a qualitative mechanistic basis for this observation.

6.4.2.1 Model Setup

The model configuration, as well as the loading and boundary conditions, are shown in Fig. 6.48. The solder joint is bonded to two copper (Cu) substrates. In the modeling horizontal displacements (Δl in the x-direction) are imposed at the far right end of the lower copper substrate. The x-direction motion of the far left edge of the upper copper is forbidden, but movement in the y-direction is allowed except that the lower-left corner of the upper copper is totally fixed. The model dimensions are: $h = 0.5$ mm, $w = 1$ mm, $H = 2.5$ mm and $L = 0.5$ mm. The relative thicknesses of

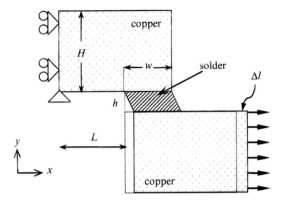

Fig. 6.48 Schematic of the solder joint model, along with the loading and boundary conditions, used in the lap-shear modeling

solder and substrate are chosen such that apparent bending resulting from the shear loading is kept at minimum. The calculations are based on the plane strain condition, which effectively simulates the nominal simple shearing mode of the solder. The nominal shear strain during deformation is defined to be $\Delta l/h$.

In the model Cu is taken to be elastic, with Young's modulus of 114 GPa and Poisson's ratio of 0.31. The Sn-based solder is assumed to be elastic-perfectly plastic, with Young's modulus 48 GPa, Poisson's ratio 0.36, and yield strength 47.9 MPa. Two forms of $LaSn_3$ particle dispersion are considered (see Fig. 6.49), one with an area fraction of 5% and the other 8%. The size and spatial distribution of particles are arbitrary. In actual materials, the particles are much smaller and more densely distributed than those adopted here. Nevertheless, the present model suffices for illustration purposes. The $LaSn_3$ particles are also elastic-perfectly plastic, with Young's modulus 64 GPa, Poisson's ratio 0.3, and yield strength 356 MPa. Note that the yield strength of the particles is much greater than that of the solder matrix.

6.4.2.2 Plastic Deformation Field

The modeling results are shown in Fig. 6.49. In Fig. 6.49a the contour plot of equivalent plastic strain in the solder for the case without intermetallic particles, when the applied nominal shear strain is 0.08, is presented. The corresponding contour plots for the cases with 5 and 8% particles are shown in Fig. 6.49b and c, respectively. In (a) two distinct plastic deformation bands inside the homogeneous solder, originating from the four corners but predominantly parallel to the interfaces, are evident. This is the unique deformation pattern under the nominal simple shear mode, as seen extensively in Chap. 5. When discrete particles with higher strength exist, Fig. 6.49b and c, the plastic flow field is seen to be disturbed to a great extent. The concentrated band becomes branched due to the blocking particles,

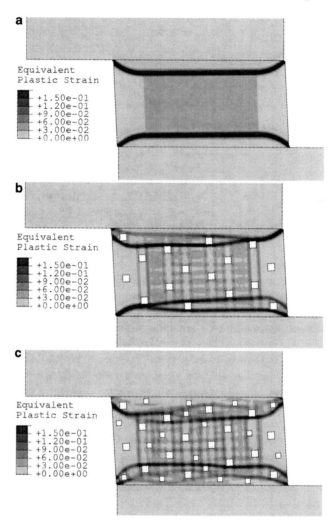

Fig. 6.49 Contour plots of equivalent plastic strain in the solder containing LaSn$_3$ intermetallic particles with area fractions of (**a**) 0%, (**b**) 5% and (**c**) 8%, when the applied nominal shear strain is at 0.08

and the maximum plastic strain is significantly reduced. This trend is more apparent in the case of more finely distributed particles (Fig. 6.49c). It can thus be expected that, if there are no other failure mechanisms active (such as debonding at the particle/matrix interface), the initiation of ductile damage in the solder matrix will be delayed. This naturally leads to an improvement of overall ductility under the same type of lap-shear loading.

The primary difference between the current example and those of particle-reinforced composites in Sect. 6.2 lies in the deformation pattern caused by the different macroscopic loading mode. In the modeling of overall uniaxial loading in Sect. 6.2, deformation is uniform in the homogeneous matrix when no particle exists; the presence of particles perturbs the deformation and induces locally higher plastic strains. In the current lap-shear loading, deformation is inherently highly localized in the homogeneous matrix. The added particles then serve to diffuse the strain concentration and result in locally lower plastic strains.

6.5 Indentation Response

Constrained deformation due to indentation loading on thin film materials attached to a thick substrate was addressed in Sect. 3.7. The focus here is on heterogeneous materials, with indentation penetration well into the composite microscopic features. This is an issue of increasing importance, because new composite materials with small-scale heterogeneities are being developed at a rapid pace, and their mechanical characterization has been, and will continue to be, dependent on the use of micro- and nanoindentation. The current status of knowledge about the indentation behavior of heterogeneous materials is quite limited. Existing theories for extracting material properties from indentation response are based on the assumption that the material being indented is a homogeneous body with a unique set of properties [94–98]. How material heterogeneity will affect the measurement and interpretation of effective material properties extracted from indentation tests is still not well understood. In addition, the complex nature of the microstructure will result in the development of a complex stress state underneath the indentation, which can lead to a non-trivial local damage state. In the presentation below we first discuss the indentation-derived yield properties of a metal-metal multilayered structure, which is followed by an extension to metal-ceramic multilayers with more detailed analyses on effective properties and local deformation field. The last section is devoted to the particle/matrix systems.

6.5.1 Metallic Multilayers: Yield Properties

Multilayered thin films consisting of alternating metallic layers have received significant attention [99–101]. Indentation hardness is very commonly used to describe their mechanical strength. What does the indentation hardness of the composite layers *really* represent? This is a notable concern. In this section we quantify the indentation hardness of the multilayers and correlate it to the overall yield property of the entire structure. It has been demonstrated that multilayers comprising of soft metals exhibit hardness over an order of magnitude higher than the rule-of-mixtures hardness of the constituents in the bulk form, which is largely due to the interface-mediated dislocation mechanisms involved in the individual layers [100, 101]. In the present study, however, we ignore the underlying defect and atomistic features and use only a continuum-based approach, for the purpose of gaining a baseline understanding of the hardness–yield strength correlation for multilayers [102]. The investigation is limited to alternating layers of two metals of equal thickness. Hypothetical elastic-plastic properties of the constituent layers are used as input in the modeling. In particular, we seek to find out if indentation hardness can provide an accurate account of the overall composite flow stress.

6.5.1.1 Model Setup

The model system is a laminated structure consisting of alternating layers of material A and material B, the same as that shown previously in Fig. 6.1. Both A and B are taken to be isotropic elastic-perfectly plastic. The Young's moduli (E) and Poisson's ratios (ν) are $E_A = 100$ GPa, $E_B = 200$ GPa, $\nu_A = 0.3$ and $\nu_B = 0.3$. The yield strengths σ_y are set to be $\sigma_{y,A} = 50$ MPa and $\sigma_{y,B} = 150$ MPa, so material A is the softer of the two. All interfaces between adjacent layers are assumed to be perfectly bonded. Note that, although the individual layer thickness may be conceived to be in the sub-macroscopic range, there is no intrinsic length scale involved in the present continuum-based simulation. As a consequence, there is no size-dependent effect caused by the varying underlying deformation mechanisms found in actual micro- and nano-layered metals. Finite element analyses of compressive loading are first performed to obtain the overall yield strength of the composite, the models of which are identical to those in Fig. 6.2a and b except that we now include the plastic properties of the layers. Both the longitudinal and transverse loading configurations are considered. The simulated composite response will be the "true" effective properties of the entire multilayer structure with all three-dimensional features accounted for.

The effective stress–strain response in the transverse direction is then used as the input response of a *homogeneous* material to be subjected to indentation loading. Since the homogeneous material used here actually represents the multilayers (with the composite material properties built in), the modeled indentation hardness will be the ideal "true" value one would seek when employing the indentation technique on the multilayer structure. However, we illustrate below that, when indentation is applied to a material with the layered structure *explicitly* accounted for, there is a significant deviation of the hardness values from the idealized "true" values.

Indentation modeling is based on the axisymmetric model featuring a rigid conical indenter, as schematically shown in Fig. 6.50. The semi-angle of the conical indenter is 70.3°, resulting in a same projected area as the Berkovich indenter [94]. There are a total of 60 alternating layers of material A and material B, all of equal thicknesses. In Fig. 6.50, the topmost layer to be in direct contact with the indenter is shown to be material A. In the analysis another model, with the topmost layer being material B, is also considered. These two multilayer arrangements are henceforth referred to as "AB stack" and "BA stack," respectively. (In the case of indenting a homogeneous material bearing the built-in composite properties, as described in the previous paragraph, the entire specimen is simply replaced by a homogeneous

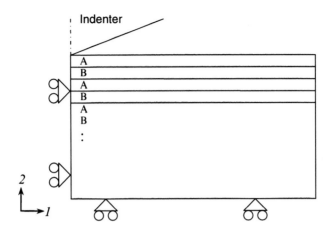

Fig. 6.50 Schematic showing the indentation model and the boundary conditions. The specimen and indenter both possess axial symmetry about the left boundary. The rigid indenter has a semi-angle of 70.3° [102]

material with the specified elastic-plastic input response.) The total thickness of the specimen is 60*t*, where *t* is the thickness of the individual layer. The lateral span (radius) of the model is 100*t*. The left boundary of the model is the symmetry axis, along which the displacement is only allowed in the vertical direction. The bottom boundary is allowed to move only in the horizontal direction. The right boundary is not constrained. The top boundary, when not in contact with the indenter, is also free to move. The coefficient of friction between the top layer and the rigid indenter is taken to be 0.1. The modeled hardness is defined to be the indentation load divided by the current projected contact area. When calculating the contact area, the last nodal point on the top surface in contact with the indenter is identified so any pile-up or sink-in effect is readily taken into consideration.

6.5.1.2 Correlating Hardness and Yield Strength

Figure 6.51 shows the modeled overall compressive stress–strain curves. Results from the longitudinal and transverse loading configurations, as well as those of pure (homogeneous) A and B used as the modeling input, are all included. It can be seen that, in both loading configurations, the multilayered composite shows a bilinear response before fully yielded. The first linear segment corresponds to a true elastic state. When the von Mises effective stress in layer A reaches $\sigma_{y,A}$, plastic yielding in A commences while B is still in the elastic state. This leads to the second linear segment. A constant flow stress ensues after both layers become plastic. This flow stress is defined to be the "composite yield strength." The composite yield strength (100 MPa) is observed to be the average of $\sigma_{y,A}$ (50 MPa) and $\sigma_{y,B}$ (150 MPa), for the present multilayer model with 50–50% volume fractions. Although transverse and longitudinal loadings both result in the same composite yield strength along the compressive axis, the individual stress components in the layers are different. Upon full yielding the algebraic values of the stress components in each layer in the transverse case (where the macroscopic compressive axis is in the 2-direction) are: $\sigma_{11}^A = -50$ MPa, $\sigma_{22}^A = -100$ MPa, $\sigma_{33}^A = -50$ MPa, $\sigma_{11}^B = 50$ MPa, $\sigma_{22}^B = -100$ MPa and $\sigma_{33}^B = 50$ MPa. In the longitudinal case (where the macroscopic compressive axis is in the 1-direction), $\sigma_{11}^A = -50$ MPa, $\sigma_{11}^B = -150$ MPa, and all other components are either zero or of very small magnitudes. One can easily check that the von Mises yield condition is satisfied in the individual materials in all these cases.

For indentation modeling, the modeled overall compressive stress–strain curve in Fig. 6.51 is used as the input response for a homogenous material, termed "homogenized multilayers" in subsequent discussion. The transverse response is chosen for this purpose, although the longitudinal response yields essentially the same indentation result since the indentation hardness under consideration is dominated by the large-deformation plastic behavior of the composite. The elastic response plays virtually no role in affecting the indentation behavior in the present analysis. Figure 6.52 shows the modeled hardness as a function of indentation depth for the pure A and B materials, the homogenized multilayers, and the composite structures with explicit A/B layers. The depth is normalized with the initial thickness of individual layers, t. It is seen that the hardness values in all homogeneous and homogenized cases are generally independent of the indentation depth. It is also observed that the model of homogenized multilayers results in hardness

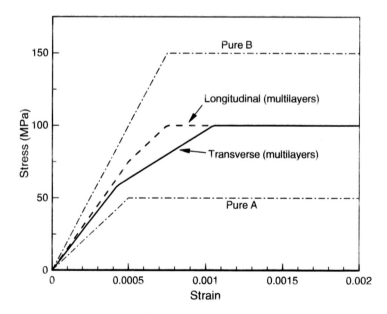

Fig. 6.51 Modeled overall compressive stress–strain curves of the multilayers. Both the longitudinal and transverse loading configurations are included. Also included for reference are the input stress–strain responses for the pure (homogeneous) A and B materials

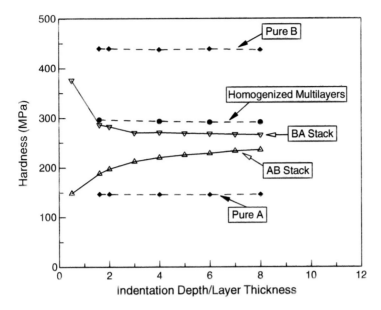

Fig. 6.52 Modeled hardness as a function of indentation depth (normalized with the initial layer thickness) for the various models considered

values which are almost exactly the averages of pure A and B materials. One can calculate the ratio of hardness/yield strength for the pure materials and homogenized multilayers shown in Fig. 6.52. Here, if the hardness values are divided by the respective yield strengths (50 MPa for "pure A," 150 MPa for "pure B," and 100 MPa for "homogenized multilayers"), the ratio is approximately 2.93 *for all three models.*

As for the composite structures in Fig. 6.52, it is evident that the "AB stack" and "BA stack" models do *not* generate the same hardness results as in the homogenized multilayers. At small depths, the hardness is dominated by the top layer material so "BA stack" and "AB stack" result in very high and low hardness values, respectively. As the indentation depth increases the difference between the two arrangements is reduced and the two curves tend to merge. Ideally there should be a single hardness value at very large indentation depths, although in Fig. 6.52 the two curves are still somewhat apart at a depth corresponding to eight initial layer thicknesses (8*t*). Nevertheless, it is apparent that both explicit composite models underestimate the overall strength of the structure (recall that the "homogenized multilayers" represents the "true" composite response). For instance, taking the average of the hardnesses of "BA stack" and "AB stack" at the maximum depth in Fig. 6.52 leads to a value of 251 MPa. Dividing this hardness value by the ratio 2.93 identified above, one obtains the *indentation-derived* composite yield strength of 85.7 MPa, which is about 14% below the "true" composite yield strength of 100 MPa. Therefore, applying indentation on the layered composite clearly leads to an *underestimation* of their overall strength.

One factor that can contribute to the underestimation of composite yield strength by indentation is the localized nature of indentation deformation. When overall deformation (such as during compressive or tensile loading) is considered, the constituent layers respond to the applied loading as well as the uniform constraint in their respective ways, so the deformation field in each layer is uniform. Under the highly localized indentation loading, such type of "ordered" behavior no longer exists. If the several softer layers close to the indentation site accommodate a greater part of the geometric constraint through easy plastic flow, an underestimation of the uniform composite strength can indeed occur. It should be noted that the present example focuses only on alternating layers of elastic-perfectly plastic films. If one or both materials strain-harden upon yielding, the indentation response may be influenced depending on the actual material parameters. Nevertheless, the analysis above serves to provide a baseline understanding of the relationship between indentation hardness and overall yield strength for multilayered elastic-plastic materials. This information is essential for further explorations involving more complex material, geometric, and interface features.

6.5.2   Metal-Ceramic Multilayers

A detailed analysis using an actual metal-ceramic multilayer system is now considered. The laminates, consisting of alternating thin layers of very different mechanical strengths, are particularly intriguing heterogeneous materials for studying indentation behavior. On the practical side, micro- and nanolayered metal-ceramic composites can possess high strength, high toughness and high damage tolerance as well as other functionalities, so they have also been a subject of intensive research [5, 7–10, 13–15, 18, 19, 21–26]. In this section we examine constrained deformation in an aluminum (Al)-silicon carbide (SiC) nanolayered composite under indentation [103].

6.5.2.1   Model Setup

Following the actual material layout used in the experiment, the model consists of 41 alternating Al and SiC layers on a substrate of silicon (Si), as shown in Fig. 6.53. Both the top layer (in contact with the indenter) and the bottom layer (adjacent to the Si substrate) are Al. A conical diamond indenter with a semi-angle of 70.3° is used. The left boundary is the symmetry axis in this axisymmetric model. The thicknesses of the individual Al and SiC layers are 50 nm each. The overall size of the entire specimen is 40 μm in lateral span (radius) and 43 μm in height. During deformation the left boundary is allowed to displace only in the 2-direction, and the bottom boundary is allowed to move only in the 1-direction. The right boundary is not constrained. As before the top layer is also free to move when not in contact with the elastic indenter. The coefficient of friction between the indenter and the top surface is 0.1.

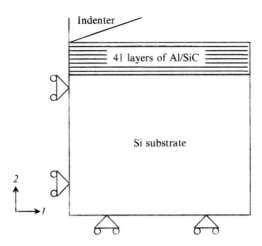

Fig. 6.53 Schematic showing the Al/SiC laminates above a Si substrate and the boundary conditions used in the axisymmetric model. The left boundary is the symmetry axis. The entire specimen is 40 μm in lateral span (radius) and 43 μm in height. The individual Al and SiC layers are 50 nm thick [103]

The Young's moduli for Al and SiC used in the model are 59 and 277 GPa, respectively. Note these values are based on actual nanoindentation measurements of single-layer Al and SiC films, and are thus different from typical values of the bulk material. The Poisson's ratios for Al and SiC are taken as 0.33 and 0.17, respectively. The plastic response of Al resembles the tensile loading data of single-layer Al, with an initial yield strength of 200 MPa. The piecewise linear, isotropic hardening response features plastic flow stresses of 300 and 400 MPa at the plastic strains of, respectively, 0.5 (50%) and 3.0 (300%), beyond which perfect plasticity is assumed. SiC is a much more brittle material. Nevertheless, a very high "yield point" of 8,770 MPa (estimated from the indentation hardness of a single-layer SiC film) is used, followed by perfect plasticity. This assumption is necessitated by the fact that a purely elastic SiC in the model will generate unrealistically high loads during indentation modeling, and is validated by the fact that in our experiment the SiC layers exhibited a glassy/plastic-type response due to the amorphous nature of the film. The Young's modulus and Poisson's ratio of the diamond indenter are 1,141 GPa and 0.07, respectively. The Young's modulus and Poisson's ratio of the Si(111) substrate are 181 GPa and 0.28, respectively. As before all the interfaces between different materials in the composite structure are assumed to be perfectly bonded.

The indentation-derived elastic modulus and hardness are directly obtained from the finite element modeling. Figure 6.54 shows the modeled load–displacement curve with a maximum indentation depth at 0.5 μm. The contact stiffness at the onset of unloading, S, can be calculated from [95]

$$S = \beta \frac{2}{\sqrt{\pi}} E_{\textit{eff}} \sqrt{A} \qquad (6.20)$$

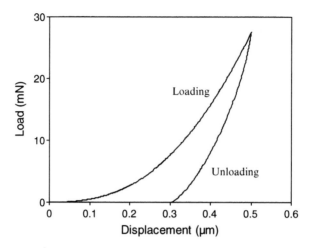

Fig. 6.54 Simulated load–displacement curve during indentation loading and unloading

with

$$\frac{1}{E_{\textit{eff}}} = \frac{1-v^2}{E} + \frac{1-v_i^{\,2}}{E_i},$$ (6.21)

where A is the projected contact area at onset of unloading, β is the indenter geometry-dependent dimensionless parameter, E and v are the Young's modulus and Poisson's ratio, respectively, of the material being indented, and E_i and v_i are the Young's modulus and Poisson's ratio, respectively, of the diamond indenter. The parameter β is first calibrated with a pure Al body of the same geometry as the entire multilayers/substrate assembly, and a value of 1.06 is determined. (This was done by making certain that the indentation-derived Young's modulus is equivalent to the input value used in the finite element modeling.) As mentioned in Sect. 6.5.1, the last nodal point on the top surface in contact with the indenter is used for calculating the projected contact area A. The determination of the composite modulus E requires a known Poisson's ratio v. Here we use a separate finite element analysis of uniaxial loading of the Al/SiC laminates to determine v, using the method depicted in Fig. 6.2b. Finally, the definition of hardness, H, is the same as before and can be calculated from

$$H = \frac{P}{A}$$ (6.22)

with P being the load at a given indentation depth and A the corresponding projected contact area.

6.5.2.2   Evolution of Stress and Deformation Fields

We now present the evolution of stress and deformation fields inside the layered structure during indentation loading and unloading. Figure 6.55 shows the hydrostatic pressure (negative hydrostatic stress) developed in the model near the indentation contact when the indentation depth is at 500 nm (part (a)) and after unloading (part (b)). The corresponding contour plots of equivalent plastic strain are shown in Fig. 6.56. The same scale is used in each group of plots so the stress and strain magnitudes can be easily compared. The deformed laminate geometry can be discerned in regions where a large shading contrast exists. It can be seen from Fig. 6.55

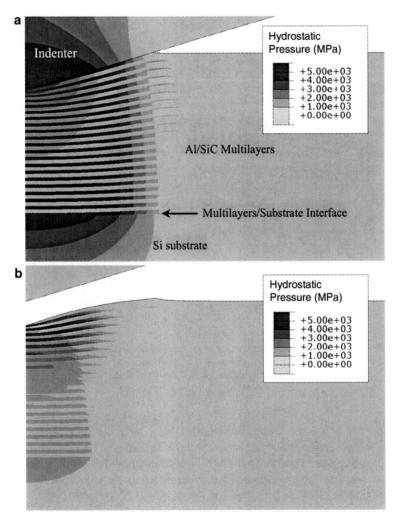

Fig. 6.55 Contour plots of hydrostatic *pressure* (negative hydrostatic stress) near the indentation site, when the indentation depth is at (**a**) 500 nm and (**b**) after unloading

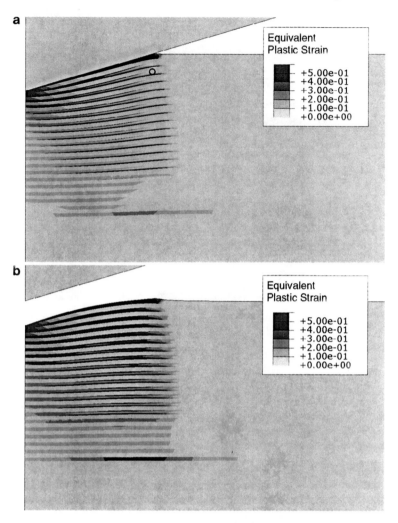

Fig. 6.56 Contour plots of equivalent plastic strain near the indentation site, when the indentation depth is at (**a**) 500 nm and (**b**) after unloading. The marking in (**a**) refers to the region where a material element in Al is selected for the tracking of stress and strain histories (see Fig. 6.57)

that, directly below the indentation contact, very large compressive hydrostatic stresses are generated. The compressive stress magnitudes in the Al layers are generally much *greater* than those in SiC. This counterintuitive result is due to the different plastic yielding response of the two materials. Note the von Mises yield condition, in terms of the principal stress components σ_1, σ_2 and σ_3, is

$$\frac{1}{\sqrt{2}}\left[\left(\sigma_1-\sigma_2\right)^2+\left(\sigma_2-\sigma_3\right)^2+\left(\sigma_3-\sigma_1\right)^2\right]^{\frac{1}{2}}=\sigma_y, \tag{6.23}$$

where σ_y is the yield strength of the material under uniaxial loading. The soft Al layers (with a small σ_y in (6.23)) have undergone severe plastic deformation, as observed in Fig. 6.56, and a large triaxial compressive stress state is developed in these layers. For the SiC layers directly below the indentation, their very high strength (σ_y in (6.23)) requires a much less compressive (or even tensile) stresses of σ_{11} and σ_{33} than the highly compressive σ_{22} (see the general form on the left-hand side of (6.23), which is dominated by the differences in the stress components). The combination of these normal stress components thus results in a much reduced hydrostatic compression in SiC because the hydrostatic stress is defined as $\frac{1}{3}(\sigma_{11} + \sigma_{22} + \sigma_{33})$. Even in the Si substrate, large hydrostatic compression exists below the indentation (Fig. 6.55). This observation suggests the importance of the substrate material in affecting the indentation response.

Upon unloading, the high stresses caused by the indentation are relieved to a great extent. However, considerably large areas, in both the multilayers and substrate, are still under residual compressive stresses well into the GPa range (Fig. 6.55b). This is a manifestation of the severity of plastic deformation occurred during loading. Another notable attribute about unloading is the evolution of deformation field in Fig. 6.56. A comparison between Fig. 6.56a and b reveals that the equivalent plastic strains in Al have actually *increased* during the *unloading* process. This observation is unique because, in a homogeneous material, indentation unloading is a pure elastic recovery process and the accumulated plastic strain remains unchanged during unloading. (Interested readers may try to confirm this by carrying out modeling on a homogeneous material. The equivalent plastic strain contours should remain exactly the same during indentation unloading.) It is apparent that, in the present case of a heterogeneous structure, the internal constraint induced by the hard and stiff SiC layers has forced the much softer Al to undergo continued deformation even though the composite as a whole is undergoing unloading.

To better understand the unloading-induced plastic deformation in Al, one can select arbitrary elements in the model and follow their deformation histories during loading and unloading. One such case is now presented. Figure 6.57a and b shows the evolution of equivalent plastic strain and several stress components (von Mises effective stress, normal stresses σ_{11}, σ_{22}, and σ_{33}, and shear stress σ_{12}), respectively, of an element inside the Al layer in a region highlighted by the open circle in Fig. 6.56a. The entire history of loading and unloading is included. In Fig. 6.57a, it can be seen that the equivalent plastic strain increases at the later stages of loading. The magnitude of plastic strain reaches about 0.08 when the indentation depth is at 0.5 μm. At the beginning of unloading, the equivalent plastic strain continues to increase and eventually attains a value of about 0.27. Note that this is *not* the "reversed yielding" effect in typical cyclic loading because there is no elastic unloading part before the restart of plastic yielding. Plastic strain here simply continues immediately upon load reversal. The plastic strain ceases to increase only after the indenter no longer makes contact with the surface.

In Fig. 6.57b, it is seen that, toward the end of indentation loading, large magnitudes of compressive stresses σ_{11}, σ_{22} and σ_{33} have developed. The shear stress σ_{12}, on the other hand, remains relatively low. The von Mises effective stress stays

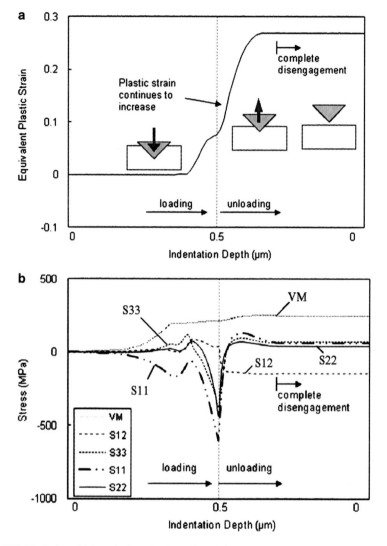

Fig. 6.57 Evolution of (**a**) equivalent plastic strain and (**b**) several stress components of a material element, inside an Al layer in the region highlighted by the *open circle* in Fig. 6.56a, during the indentation history. The symbols VM, S12, S33, S11 and S22 in (**b**) refer to the von Mises effective stress and the stress components σ_{12}, σ_{33}, σ_{11} and σ_{22}, respectively

relatively constant (it continues to increase slightly following the strain hardening response of Al). Upon unloading, the magnitudes of the normal stress components decrease rapidly. However, the magnitude of shear stress σ_{12} quickly becomes significant. The net effect is that the von Mises effective stress remains at the yielding level so plastic deformation continues. The buildup of the shear stress during unloading may be attributed to the uneven elastic relief of the adjacent SiC layers directly above and below.

The deformation history in other material elements may be tracked in a similar manner. The continuation of plastic deformation in Al during unloading is a consequence of the internal mechanical constraint imposed by the hard ceramic layers. The present finding does raise the issue about the accuracy of using instrumented indentation, during unloading, to measure the elastic behavior of multilayered materials, because the unloading process is no longer a simple elastic event.

Figure 6.58 shows the contour plots of σ_{22} when the indentation depth is at (a) 500 nm and (b) after unloading. The contour levels are adjusted so as to highlight the *tensile* σ_{22} stresses evolved during loading and unloading (otherwise the localized

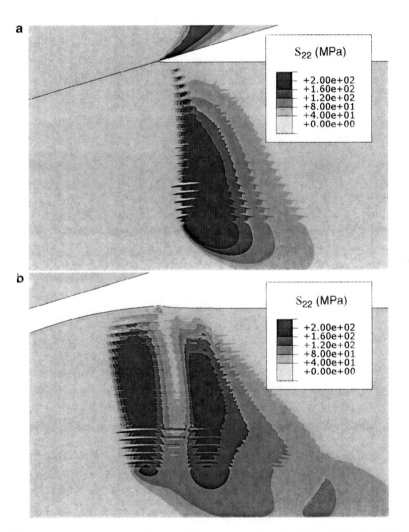

Fig. 6.58 Contour plots of σ_{22} near the indentation site, when the indentation depth is at (a) 500 nm and (b) after unloading. The contour levels are adjusted to highlight the region with tensile stresses

tensile stresses would be largely "hidden" by the predominant compressive field). It can be seen that, under indentation, tensile stresses in the vertical direction have developed in certain areas below and slightly outside the edge of indentation. The stress magnitude can be significant, well over 200 MPa in many Al and SiC layers as well as in a small region of the Si substrate (Fig. 6.58a). Upon unloading, Fig. 6.58b, the region with high tensile stresses has shifted slightly inward and two distinct high-stress areas have developed. It is noticed, by comparing the lighter shades in Fig. 6.58a and b, that the area with tensile σ_{22} stresses has actually *expanded* during *unloading*. The expansion is due to the fact that the heavily compressed material directly below the indentation contact has undergone elastic recovery during unloading. This imposes a vertical pulling action on the region already under tension. In fact, the same qualitative features can be seen even in a homogeneous material, and they are now enhanced by the presence of the laminated structure. Experimental characterization on post-indented Al/SiC nanolayers have revealed local cracking and delamination, below and slightly outside the edge of indentation, in mid-layers and along the interface between the multilayers and Si substrate [103]. It is notable that these indentation-induced damages fall into the region where large tensile σ_{22} stresses are seen in the modeling.

6.5.2.3   Overall Indentation Response

The overall hardness and elastic modulus of the multilayers obtained from indentation modeling are now considered. For the purpose of gaining fundamental insight into the indentation response, two additional models are used as depicted in Fig. 6.59a and b. First, the 41 layers of Al/SiC in Fig. 6.53 are replaced by a homogeneous material having the effective properties of the Al/SiC multilayered composite, Fig. 6.59a. The Si substrate in the model remains unchanged. The effective elastic-plastic stress–strain response of the homogenized material is obtained from a separate finite element analysis of overall compression loading along the through-thickness direction (Fig. 6.2b). In particular, the anisotropic elastic behavior, with five independent elastic constants for the transversely isotropic composite, is accounted for in the homogenized model (Sect. 6.1.1). In Fig. 6.59b, the Si substrate is also replaced by the homogenized Al/SiC multilayers. The difference between the cases in Fig. 6.59a and b is that the effect of the Si substrate on the indentation response can be examined. The models in Figs. 6.53, 6.59a and b are henceforth termed "Al/SiC multilayers on Si," "homogenized Al/SiC on Si," and "homogenized Al/SiC," respectively.

Figure 6.60 shows the modeled indentation hardness as a function of indentation depth for the three models, up to a maximum depth of 1,000 nm (about one half of the total thickness of the multilayers). It can be seen, by comparing the cases of "homogenized Al/SiC on Si" and "homogenized Al/SiC," that the Si substrate effect on hardness is negligible when the indentation is shallower than about 500 nm. The influence of the substrate becomes apparent at a depth of 1,000 nm, where approximately a 12% increase in hardness is observed. The model "Al/SiC

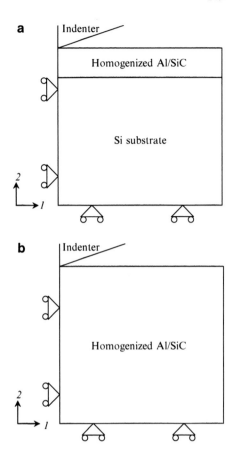

Fig. 6.59 Schematics showing the two additional models: (**a**) the 41 layers of Al/SiC is replaced by the homogenized Al/SiC composite above the Si substrate; (**b**) the entire material is the homogenized Al/SiC composite [103]

multilayers on Si" in Fig. 6.60, however, shows much lower hardness values compared to the other two. At small depths, the top Al layer plays a relatively important role in affecting the composite modulus so a lower hardness results. The overall hardness increases as the indentation goes deeper (due to the increasing influence of SiC layers as well as the Si substrate). However, at 1,000 nm the hardness is still quite far behind the homogenized composite response. Note that the general trend observed here is consistent with that of the metal-metal multilayers in Sect. 6.5.1: the model with explicit layers ("Al/SiC Multilayers on Si") tends to underestimate the hardness of the homogenized model ("Homogenized Al/SiC on Si").

It is worth mentioning that the difference between the model with explicit multilayers and that of homogenized composite, shown in Fig. 6.60, is not merely a consequence of the indentation depth considered in the analysis. With the homogenized approach, the pile-up at the indentation edge is found to be significantly

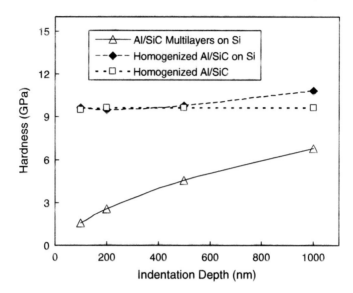

Fig. 6.60 Simulated indentation hardness as a function of indentation depth, for the three models "Al/SiC multilayers on Si" (Fig. 6.53), "homogenized Al/SiC on Si" (Fig. 6.59a), and "homogenized Al/SiC" (Fig. 6.59b)

reduced compared to the case with the individual Al/SiC multilayers. As a consequence the projected contact area, A, used in obtaining the hardness, is also significantly affected. This fundamental difference, along with the discrepancy shown in Fig. 6.60, illustrate that, when modeling the indentation behavior of a heterogeneous structure such as the multilayers considered here, a homogenization approach may yield very inaccurate results.

Figure 6.61 shows the modeled elastic modulus, obtained from the indentation unloading response, as a function of indentation depth for the three models, "Al/SiC multilayers on Si," "homogenized Al/SiC on Si," and "homogenized Al/SiC." A comparison between the cases of "homogenized Al/SiC on Si" and "homogenized Al/SiC" reveals the substrate effect. Note the Young's modulus of Si, 181 GPa, is still much greater than the effective modulus of the Al/SiC composite along the thickness direction (about 117 GPa), so an increasing contribution from Si is seen in the model due to the increasing substrate effect. From Figs. 6.60 and 6.61 it can be seen that the effect of substrate on the indentation-derived modulus is much more pronounced than on the hardness. (A similar relation was observed in Sect. 3.7.) The substrate effect on the elastic modulus is significant even when the indentation is shallow (e.g., well within 10% of the homogenized multilayered film).

When the explicit Al and SiC layers are included in the model ("Al/SiC multilayers on Si"), however, a different behavior of the indentation-derived elastic modulus is observed as in Fig. 6.61. Here the modulus remains relatively constant throughout the range of indentation depth. The curve deviates from that of the "homogenized Al/SiC on Si" model as the indentation depth increases, and the apparent substrate effect

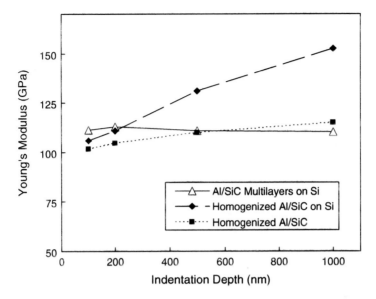

Fig. 6.61 Simulated elastic modulus as a function of indentation depth obtained from the simulated unloading part of the indentation curve, for the three models "Al/SiC multilayers on Si" (Fig. 6.53), "homogenized Al/SiC on Si" (Fig. 6.59a), and "homogenized Al/SiC" (Fig. 6.59b)

no longer exists. This may be attributed to the extensive plastic deformation in Al during unloading as discussed above. The continuation of plastic deformation, leading to a reduced contact stiffness, can offset the substrate effect, and this behavior becomes increasingly significant as the indentation depth increases. The result in Fig. 6.61 again raises the issue about the accuracy of using instrumented indentation to quantify the elastic behavior of multilayered composites.

6.5.3 Particle-Matrix Systems

We now return to composite materials consisting of a soft matrix and hard particles, but with a focus on indentation loading. Indentation on this type of heterogeneous materials is more difficult to model, because the axisymmetric approach can no longer be adopted (otherwise each particle away from the symmetry axis becomes ring-like physically). Using a 3D model, with sufficient numbers of particles distributed over a volume with sufficient depth, may be prohibitively costly, in terms of the number of finite elements that has to be included to resolve the internal geometric details. For gaining qualitative understanding and insight, one can resort to 2D planar models [104, 105]. In this section we present a numerical example of ceramic particle-reinforced metal matrix composite. Attention is again devoted to testing the validity of the "homogenization" approach applied to indentation modeling.

6.5.3.1 Model Setup

A schematic depicting the modeling approach is shown in Fig. 6.62. The computational domain is a square containing distributed square-shaped particles. The particles are elastic, and the matrix is a ductile elastic-plastic phase. The finite element modeling consists of two steps. A uniform stretching is first simulated to obtain the overall stress–strain curve of the composite. This stress–strain relation then serves as the input response for a homogeneous material, termed "homogenized composite" in Fig. 6.62, which is then subject to indentation loading. Indentation is simulated by pressing a rigid circular indenter onto the top surface. In a parallel manner indentation is also simulated directly on the two-phase (particle-bearing) composite. The significance of this approach is that the two systems under indentation (i.e., homogenized composite and two-phase composite) have exactly the same *overall* stress–strain response, and we seek to find out if the modeling can lead to the same indentation response [104].

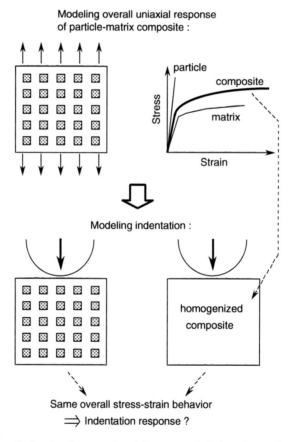

Fig. 6.62 Schematic showing the general modeling approach. Indentation on the two-phase composite and the homogenized composite is simulated. The homogenized composite is taken to have the same overall stress–strain characteristics as the two-phase composite [104]

In the simulation of tensile loading the problem pertains to the common unit-cell approach (Sects. 6.1 and 6.2). For indentation, only the right half of the structure is used because of symmetry (see Fig. 6.65 below). The bottom boundary is allowed to slip tangentially during indentation. The top boundary is not constrained, except that a coefficient of friction of 1.0 is imposed when a contact with the rigid indenter is established. The left boundary is the symmetry axis, and the right boundary is free to move but is constrained to remain vertical. (Setting the right boundary to be entirely free produces only slightly different indentation response.) Various total numbers of particles are used, and the case of 144 particles is treated as the standard case for the presentation below. The area fraction of particles is taken to be 0.16. The ratio of indenter radius and initial side length of the specimen is taken to be 0.79.

The analyses are based on the plane stress formulation. A plane strain model would cause complexities when carrying the simulated overall stress–strain response to the homogenized material model, due to the constraint in the third direction. Therefore we choose to use plane stress, which, despite its idealized sheet-like nature, is capable of representing a two-dimensional composite in the present case. Salient qualitative features of composite strengthening and local deformation pattern can be adequately accounted for without ambiguity.

Young's modulus and Poisson's ratio of the particles are 450 GPa and 0.17, respectively. The matrix phase has Young's modulus 70 GPa and Poisson's ratio 0.33, with its plastic portion following a piecewise linear hardening response. Two cases of matrix strength in the plastic regime are considered. In the "strong" matrix case, the plastic response is taken to have flow stress values of 360, 450, 525 and 750 MPa at the plastic strains of 0, 0.02, 0.05 and 0.50, respectively. In the "weak" matrix case, the stress levels are two third of those in the "strong" case at the same plastic strains. The uniaxial stress–strain curves for the two matrices as well as the particles are shown in Fig. 6.63. As a consequence, two sets of input parameters, resulting from the "strong" and "weak" cases, are fed into the homogenized composite model.

6.5.3.2 Indentation Response

Figure 6.64 shows the modeled response of load versus displacement during indentation, for both the cases of "strong" and "weak" matrices. The indentation displacement is normalized, defined to be the ratio of the penetration depth and the initial height of the specimen. Note that the indentation response curves downward, which is opposite to that in the axisymmetric model (and actual experiment) shown in Figs. 6.54 and 3.14. This is due to the planar nature of the present model. It is evident in Fig. 6.64 that a stronger matrix requires a greater load to achieve the same indentation depth. A more important issue at hand, however, is the comparison of results between the cases of two-phase and homogenized composites. After an initial stage up to a displacement of about 0.03, the two-phase model clearly displays a harder response than the homogenized model. At a displacement of 0.15, the difference in load reaches about 12% in both the cases of "strong" and "weak" matrices. This is an

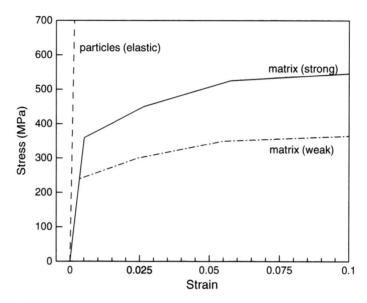

Fig. 6.63 Uniaxial stress–strain response of the particles, the "strong" matrix and the "weak" matrix used in the modeling

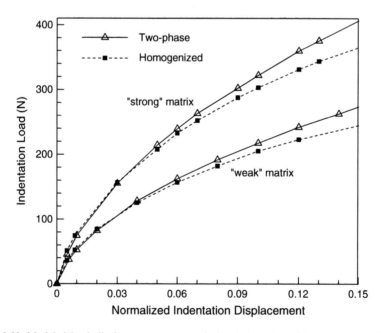

Fig. 6.64 Modeled load–displacement response during indentation of both the "strong" and "weak" matrix composites. The displacement is normalized by the initial specimen height

indication that treating the particle-containing material as a homogeneous continuum leads to a different indentation result. When the particles are explicitly included, a significantly harder response is obtained irrespective of the fact that the two-phase and homogenized composites have the same overall stress-strain response.

Figure 6.65a and b show the contour plots of equivalent plastic strain in the homogenized and two-phase composites, respectively, at an indentation depth of 0.10. These plots correspond to the "strong" matrix case; the "weak" matrix case results in qualitatively the same features. In Fig. 6.65b most of the particles are well discerned because no plasticity exists in the purely elastic particles. It can be seen that, in both cases, strong plasticity is localized, and the strain magnitude is significantly reduced in regions away from the indentation site. The deformation patterns, however, are very different in Fig. 6.65a and b. The presence of particles dramatically alters the plastic strain field and drives the deformation into a banded structure. As a consequence, different indentation response resulting from the two models can be expected. Figure 6.65b also illustrates that the particles directly below the depression are displaced downward with the surrounding matrix.

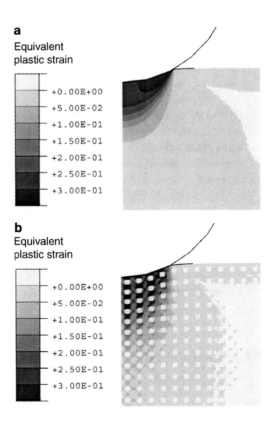

Fig. 6.65 Contour plots of equivalent plastic strain in the (**a**) homogenized and (**b**) two-phase composites when the normalized indentation displacement is at 0.10

In fact, the local density of particles is forced to increase, compared to other locations. This can be appreciated by noting that the inter-particle spacing, especially along the vertical direction, is reduced in the region directly below the indentation. As the indenter travels downward, it encounters resistance from a material with an increasingly greater concentration of hard particles. This increase in particle concentration also renders the locally strain-hardened matrix even more constrained. A greater load is thus needed to attain a given displacement. In the homogenized composite, this hardening effect due to "particle crowding" is nonexistent. The general trend observed in Fig. 6.64 can thus be rationalized.

Although the two-phase composite shows a generally harder indentation response, at the very early stage of indentation (displacement smaller than about 0.03), however, the homogenized composite is actually slightly harder (Fig. 6.64). This is because that, when the depression is still very shallow in the two-phase composite, the indenter does not "feel" the influence of particles and the resistance to indentation arises mainly from the soft matrix. In the homogenized model, however, the indenter immediately encounters a homogeneous material harder than the soft matrix (i.e., with the strengthening effects already built-in). Consequently, a harder indentation response is seen for the homogenized model at the beginning. As the indenter moves deeper, the particle effect becomes dominant in the two-phase model so the trend is reversed thereafter.

The above features are also observed in simulations incorporating different numbers and more random distributions of particles. The discrepancy resulting from the two approaches (two-phase versus homogenized) persists, since the local increase of particle concentration cannot be accounted for in the homogenized model. At this juncture one question regarding the two-phase model may arise: are enough particles included in the analysis so the indentation behavior may be deemed representative of the composite with very many particles? To address the question one can examine the convergence of numerical results by varying the total number of particles under a fixed particle concentration. Figure 6.66 shows such a comparison of indentation load-displacement response. The area fraction of particles is 0.16 in all cases and the number of particles included in the model is increased from 1 to 144. All particles are regularly arrayed, with the one-particle case having a single large particle at the center of the square computational domain. The softer response of the one-particle case is evident, due to the fact that the soft matrix near the top makes the most contribution in resisting the indentation. As the number of particles increases, however, the indentation response soon converges. The curves for the cases of 100 and 144 particles show very little difference. Therefore, it is reasonable to conclude that the case of 144 particles used for the presentation above is representative of the "many particles" scenario.

6.5.3.3 Implications

The present example clearly illustrates that, in modeling the indentation behavior of particle-matrix type of composites, one should exercise caution in endowing the

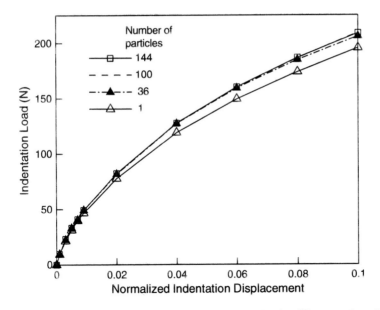

Fig. 6.66 Modeled load–displacement response during indentation for different total numbers of particles included. The area fraction of particles is fixed at 0.16

composite with a simple constitutive response because it may not be a logical approach, regardless of the length scales involved. The actual indentation response (e.g., from the "two-phase" approach) may not be well represented by treating the material as a simple continuum (such as the "homogenized" approach). This problem is largely nonexistent in modeling the overall stress–strain response, owing to the fundamental difference between the two loading modes as discussed in previous sections on multilayered composites. During a tensile (or compressive) test, the material within the gauge section undergoes nominally uniform deformation. In an indentation test, however, severe plastic deformation is concentrated in a confined region directly below the indentation, outside of which the material still behaves elastically. The density of particles is forced to increase directly below the indentation. To this end, it is worth mentioning that, in experimental studies using SiC particle-reinforced Al matrix composites [106, 107], the effects of this "particle crowding" in influencing the macro-hardness behavior have been observed and evaluated.

As a final remark, we recognize an interesting contrast between the cases of particle-reinforced composite and multilayer composite. In Fig. 6.64, the use of an explicit composite structure (two-phase) in the model gives rise to a *harder* indentation response than the homogenized counterpart. On the other hand in Sects. 6.5.1 and 6.5.2, specifically Figs. 6.52 and 6.60, the use of explicit multilayers in the model is seen to predict *lower hardness* compared to the homogenized approach. The actual form of heterogeneity apparently plays a key role in dictating the fundamental indentation behavior of the material.

6.6   Projects

1. In Sect. 6.1.1 the effective elastic property of multilayers was discussed. There are five independent elastic constants for the transversely isotropic structure. In the numerical example the Young's modulus E and Poisson's ratio v of the constituents A and B were: $E_A = 59$ GPa, $v_A = 0.33$, $E_B = 277$ GPa and $v_B = 0.17$. Obtain the five independent effective elastic constants for the multilayers with equal thicknesses of A and B, following the approach outlined in the section. If the Poisson's ratios of A and B are made equivalent (say, 0.33 or 0.17 for both), how different will the result be? In addition, explore the effect of Poisson's ratio by making it even larger and smaller.

2. Equation (6.7) gives the rule-of-mixtures expression of the longitudinal elastic modulus of long-fiber composites. Construct a finite element model with a periodic fiber array (say the hexagonal type), and simulate loading along the fiber direction to obtain the longitudinal modulus. Use various sets of fiber and matrix properties and quantify the errors given by (6.7).

3. It was shown in Sect. 6.1.2 that the hexagonal fiber array in a long-fiber composite results in transverse isotropy for elastic behavior. What about plastic behavior? Conduct a modeling analysis to find out, using the generalized plane strain unit-cell models depicted in Fig. 6.4. What if the matrix material is viscoelastic?

4. For long-fiber composites with a periodic hexagonal fiber arrangement, the material is transversely isotropic during elastic deformation (Sect. 6.1.2). Devise a numerical strategy to obtain all five independent elastic constants, similar (in spirit) to the procedures in Sect. 6.1.1 for the multilayered composite.

5. The effects of shape and spatial distribution of particles on the overall plastic behavior of SiC particle-reinforced Al matrix composites were discussed in Sects. 6.2.1 and 6.2.2. Conduct a systematic analysis using a viscoplastic Al matrix. (You may do a literature search to identify an appropriate viscoplastic model for Al.) Examine the additional effects of applied strain rate.

6. The problem is similar to Project 5 above, except that the Al matrix now features a creep constitutive response. (You may do a literature search to identify an appropriate creep model for Al.) Study the creep and stress relaxation behaviors, combined with the particle shape and distribution effects, of the composite.

7. Consider the effect of particle distribution on the overall plastic response of particle-reinforced composites discussed in Sect. 6.2.2 (specifically, the square-type arrangements in Fig. 6.15). Perform numerical modeling using model B without the small particles. Check if your composite stress–strain curve and local deformation field are equivalent to those with the small particles (as shown in Figs. 6.16 and 6.17). In addition, conduct an in-depth study by designing different composite structures with various size and spatial distributions of particles. Focus on the correlation between local stress and strain fields and the macroscopic stress–strain behavior obtained from your models.

8. Consider the effect of thermal residual stresses on the overall elastic-plastic response of metal matrix composites discussed in Sect. 6.2.3. Conduct a series

of modeling analysis with the same approach, using various reinforcement shapes and volume fractions as well as different yield behavior (perfectly plastic, different extents of matrix yield strength and strain hardening etc.). What about compressive loading instead of tensile loading after the cooling step? The examples given in Sect. 6.2.3 were based on the assumption of isotropic hardening of the matrix material. Will there be any difference if you utilize kinematic hardening?

9. Consider the effect of thermal residual stresses on the overall elastic-plastic response of metal matrix composites discussed in Sect. 6.2.3. Perform a systematic modeling analysis focusing on the influence of spatial arrangement of the reinforcing particles. You may use the 2D plane strain approach and the geometric models depicted in Figs. 6.15 and 6.18. Include in your study various particle volume fractions and matrix yield properties (see also Project 8). What about compressive loading instead of tensile loading after the cooling step? Will the isotropic hardening and kinematic hardening models for the matrix material produce different results?

10. With reference to the cyclic response of the particle-reinforced metal matrix composite considered in Sect. 6.2.4, design and conduct a series of modeling analysis featuring different reinforcement shapes and volume fractions. Use several strain ranges. Use also strain ranges biased toward the tensile side (e.g., between −0.005 and 0.025 etc.) and compressive side. You may also explore stress-controlled cyclic response by imposing different stress ranges. Study how kinematic hardening and isotropic hardening will lead to different macroscopic behavior. Correlate the macroscopic stress–strain curves with the microscopic (local) stress and deformation fields.

11. Design a 3D numerical study following the approach in Sect. 6.3.2 on the thermal expansion behavior of metal-matrix, ceramic-matrix and interpenetrating composites. However, use kinematic hardening for the plastic yielding behavior of metal instead of isotropic hardening. Will your results show any qualitative differences? You may also perform thermal cycling modeling, and correlate the macroscopic strain–temperature response with the evolution of local stress and deformation fields.

12. Design a 3D numerical study of mechanical loading using the 3D model configurations in Sect. 6.3.2 (metal-matrix, ceramic-matrix and interpenetrating composites). Use your 3D models to study the elastic, plastic and/or time-dependent creep behaviors of the three classes of composites.

13. Section 6.3.3 presents a series of analysis on the thermal expansion behavior of metal-ceramic composites containing a small volume fraction of voids. The Al matrix was assumed to follow the elastic-plastic response with isotropic hardening. Carry out similar modeling analyses using the kinematic hardening model. Will the strain–temperature responses of the composite look fundamentally different? Also, design 2D models with different types of phase contiguity, void configuration, void concentration, particle shape and/or particle contact, and study their influences on the composite thermal expansion.

14. With reference to the modeling of CTE of the epoxy matrix/silica filler composites in Sect. 6.3.4, conduct a systematic study using different viscoelastic parameters for the epoxy. (You may consult the literature regarding the choice of viscoelastic model and parameters.) Examine the local stress and deformation fields in detail, and correlate them with the macroscopic thermal expansion behavior. Explore the effects of a wide range of heating and cooling rates, as well as the stress relaxation response of the composite.

15. In Sect. 6.4.2 the addition of particles to the solder matrix is seen to be able to reduce the plastic flow concentration and strain magnitude. Conduct a series of modeling analysis using models incorporating more and smaller particles (without increasing the overall particle concentration). Examine the plastic strain field as well as the overall shear stress–shear strain response. If you include the viscoplastic behavior of the Sn matrix, how will the applied strain rate affect the results?

16. With reference to the indentation analysis of metal-metal multilayers presented in Sect. 6.5.1, conduct similar analyses using different friction conditions between the indenter and the top layer. How will the result be quantitatively affected? What if you change the yield strengths of materials A and B? You may also try to include certain forms of strain hardening and perform a systematic study.

17. In Sect. 6.5.2 it was shown that, during indentation unloading, plastic deformation in the Al layers can continue to occur in the Al/SiC multilayered composite. Conduct a similar modeling analysis. Choose several other material elements in both the Al and SiC layers, and track their stress and strain histories during loading and unloading. Correlate these local responses with the overall loading and discuss their significance. You may also devise models with different relative thicknesses and/or strengths of the metal and ceramic, and study their effects on the "unloading plasticity" in a systematic way.

18. In Sect. 6.5.2, the indentation-derived properties of the Al/SiC multilayers were seen to be influenced by the Si substrate, especially for the elastic modulus. One way to eliminate the influence of the substrate is to use a multilayer model containing a great number of layers (hundreds or beyond) to represent a "true" multilayer response. But this is impractical in that the number of finite elements will be prohibitively large to resolve the layered geometry throughout the multilayer domain. A more practical way is to include a reasonable number of explicit layers (e.g., within 100) on top, and use a substrate having the same properties of the *homogenized* multilayers (so the finite element mesh can be made coarser in the substrate region). With reference to Fig. 6.53, for instance, the Si substrate can be replaced with the "homogenized Al/SiC." The model then becomes one of essentially "infinite" layers, but the distinct layered features near the indentation still remain. Construct such a model and carry out the indentation modeling. Examine the evolution of stress and deformation fields, as well as the indentation-derived hardness and elastic modulus as a function of indentation depth. Compare your results with those other cases as in Sect. 6.5.2.

19. With reference to the continuation of plastic deformation in the metal-ceramic multilayers during indentation unloading presented in Sect. 6.5.2, conduct a

similar type of indentation modeling but using a hard particle/soft matrix composite material. Here the axisymmetric model is no longer feasible and a 2D plane-strain model should suffice. Can you observe any plastic deformation during unloading? Are you able to identify other heterogeneous geometries that undergo plasticity upon unloading? Explore your ideas by way of finite element modeling.

20. In Sect. 6.5.1 (Fig. 6.52) and Sect. 6.5.2 (Fig. 6.60), the use of the explicit multilayered structure in indentation modeling was seen to predict lower hardness values than the homogenized model. However, in Sect. 6.5.3 (Fig. 6.64) an explicit particle-containing composite model was seen to predict a harder response. Can you think of a reason for these seemingly opposite trends? What might be the rules of material heterogeneity that govern the trend (upward or downward, compared to the homogenized case)? Design different numerical models to aid in your exploration.

21. Indentation modeling of the particle-matrix composite system was considered in Sect. 6.5.3. If the particles are made to be voids, the model then becomes one of porous materials. One can imagine that, during indentation, pore-crushing takes place, which will affect the correlation between the indentation response and overall stress–strain behavior of the material. Design a porous model and carry out a systematic study following the approach in Sect. 6.5.3. If your computing hardware capability allows, perform 3D simulations. This is a technologically important issue, because porous thin films have been, and are being, developed for a variety of applications. Nanoindentation is commonly used to characterize their mechanical properties (modulus and strength etc.). But what is the true relationship between the indentation-derived properties and the true overall mechanical properties for these porous materials?

References

1. R. M. Christensen (1979) Mechanics of composite materials, Wiley, New York.
2. R. M. Jones (1999) Mechanics of composite materials, 2nd ed., Taylor and Francis, Philadelphia.
3. Z. Hashin (1983) "Analysis of composite materials – A survey," Journal of Applied Mechanics, vol. 50, pp. 481–505.
4. K. K. Chawla (1998) Composite Materials, Springer, 2nd ed., New York.
5. A. T. Alpas, J. D. Embury, D. A. Hardwick and R. W. Springer (1990) "The mechanical properties of laminated microscale composites of Al/Al$_2$O$_3$," Journal of Materials Science, vol. 25, pp. 1603–1609.
6. D. O. Northwood and A. T. Alpas (1998) "Mechanical and tribological properties of nanocrystalline and nanolaminated surface coatings," Nanostructured Materials, vol. 10, pp. 777–793.
7. G. T. Mearini and R. W. Hoffman (1993) "Tensile properties of aluminum/alumina multilayered thin films," Journal of Electronic Materials, vol. 22, pp. 623–629.
8. T. C. Chou, T. G. Niwh, T. Y. Tsui, G. M. Pharr and W. C. Oliver (1992) "Mechanical properties and microstructures of metal/ceramic microlaminates: Part I. Nb/MoSi2 systems," Journal of Materials Research, vol. 7, pp. 2765–2773.
9. T. C. Chou, T. G. Nieh, S. D. McAdams, G. M. Pharr and W. C. Oliver (1992) "Mechanical properties and microstructures of metal/ceramic microlaminates: Part II. A Mo/Al$_2$O$_3$ system," Journal of Materials Research, vol. 7, pp. 2774–2784.

10. C. H. Liu, Wen-Zhi Li and Heng-De Li (1996) "TiC/metal nacreous structures and their fracture toughness increase," Journal of Materials Research, vol. 11, pp. 2231–2235.
11. V. P. Godbole, K. Dovidenko, A. K. Sharma and J. Narayan (1999) "Thermal reactions and micro-structure of TiN–AlN layered nano-composites," Materials Science and Engineering B, vol. 68, pp. 85–90.
12. P. C. LeBaron, Z. Wang and T. J. Pinnavaia (1999) "Polymer-layered silicate nanocomposites: an overview," Applied Clay Science, vol. 15, pp. 11–29.
13. M. Ben Daia, P. Aubert, S. Labdi, C. Sant, F. A. Sadi, Ph. Houdy and J. L. Bozet (2000) "Nanoindentation investigation of Ti/TiN multilayers films," Journal of Applied Physics, vol. 87, pp. 7753–7757.
14. J. H. Lee, W. M. Kim, T. S. Lee, M. K. Chung, B.-K. Cheong and S. G. Kim (2000) "Mechanical and adhesion properties of Al/AlN multilayered thin films," Surface and Coatings Technology, vol. 133–134, pp. 220–226.
15. A. Lousa, J. Romero , E. Martýnez , J. Esteve , F. Montala and L. Carreras (2001) "Multilayered chromium/chromium nitride coatings for use in pressure die-casting," Surface and Coatings Technology, vol. 146–147, pp. 268–273.
16. I. Luzinov, D. Julthongpiput, V. Gorbunov and V. V. Tsukruk (2001) "Nanotribological behavior of tethered reinforced polymer nanolayer coatings," Tribology International, vol. 34, pp. 327–333.
17. M. Xiao, L. Sun, J. Liu, Y. Li and K. Gong (2002) "Synthesis and properties of polystyrene/graphite nanocomposites," Polymer, vol. 43, pp. 2245–2248.
18. J. Romero, A. Lousa, E. Martinez and J. Esteve (2003) "Nanometric chromium/chromium carbide multilayers for tribological applications," Surface and Coatings Technology, vol. 163–164, pp. 392–397.
19. M. A. Phillips, B. M. Clemens and W. D. Nix (2003) "Microstructure and nanoindentation hardness of Al/Al$_3$Sc multilayers," Acta Materialia, vol. 51, pp. 3171–3184.
20. D.-H. Kuo and K.-H. Tzeng (2004) "Characterization and properties of r.f.-sputtered thin films of the alumina-titania system," Thin Solid Films, vol. 460, pp. 327–334.
21. X. Deng, N. Chawla, K. K. Chawla, M. Koopman and J. P. Chu (2005) "Mechanical behavior of multilayered nanoscale metal-ceramic composites," Advanced Engineering Materials, vol. 7, pp. 1099–1108.
22. N. Chawla, D. R. P. Singh, Y.-L. Shen, G. Tang and K. K. Chawla (2008) "Indentation mechanics and fracture behavior of metal/ceramic nanolaminate composites," Journal of Materials Science, vol. 43, pp. 4383–4390.
23. D. Bhattacharyya, N. A. Mara, R. G. Hoagland and A. Misra (2008) "Nanoindentation and microstructural studies of Al/TiN multilayers with unequal volume fractions," Scripta Materialia, vol. 58, pp. 981–984.
24. G. Tang, D. R. P. Singh, Y.-L. Shen and N. Chawla (2009) "Elastic properties of metal-ceramic nanolaminates measured by nanoindentation," Materials Science and Engineering A, vol. 502, pp. 79–84.
25. D.L. Windt and J.A. Bellotti (2009) "Performance, structure, and stability of SiC/Al multilayer films for extreme ultraviolet applications," Applied Optics, vol. 48, pp. 4932–4941.
26. P. Jonnard, K. Le Guen, M.-H. Hu, J.-M. André, E. Meltchakov, C. Hecquet, F. Delmotte and A. Galtayries (2009) "Optical, chemical and depth characterization of Al/SiC periodic multilayers," Proceedings of SPIE, vol. 7360, 73600O.
27. G. Tang, Y.-L. Shen, D. R. P. Singh and N. Chawla (2008) "Analysis of indentation-derived effective elastic modulus of metal-ceramic multilayers," International Journal of Mechanics and Materials in Design, vol. 4, pp. 391–398.
28. D. Hull and T. W. Clyne (1996) An introduction to composite materials, 2nd ed., Cambridge University Press, Cambridge.
29. N. Chawla and K. K. Chawla (2006) Metal matrix composites, Springer, New York.
30. S. Suresh and A. Mortensen (1998) Fundamentals of functionally graded materials – processing and thermomechanical behavior of graded metals and metal-ceramic composites, IOM Communications, London.

31. T. Nakamura and S. Suresh (1993) "Effects of thermal residual stresses and fiber packing on deformation of metal-matrix composites," Acta Metallurgica et Materialia, vol. 41, pp. 1665–1681.

32. J. R. Brockenbrough, S. Suresh and H. A. Wienecke (1991) "Deformation of metal-matrix composites with continuous fibers: Geometrical effects of fiber distribution and shape," Acta Metallurgica et Materialia, vol. 39, pp. 735–752.

33. J. L. Teply and G. J. Dvorak (1988) "Bounds on overall instantaneous properties of elastic-plastic composites," Journal of the Mechanics and Physics of Solids, vol. 36, pp. 29–58.

34. H. J. Bohm and F. G. Rammerstorfer (1991) "Micromechanical investigation of the processing and loading of fiber-reinforced metal matrix composites," Materials Science and Engineering A, vol. 135, pp. 185–188.

35. S. Nemat-Nasser and M. Hori (1993) Micromechanics: overall properties of heterogeneous materials, North-Holland, Amsterdam.

36. R. K. Everett and R. J. Arsenault (1991) Metal matrix composites: mechanisms and properties, Academic Press, Boston.

37. D. Francois, A. Pineau and A. Zaoui (1998) Mechanical behavior of materials, Volume I: elasticity and plasticity, Kluwer, Dordrecht.

38. Z. Hashin and S. Shtrikman (1963) "A variational approach to the theory of elastic behavior of multiphase materials," Journal of the Mechanics and Physics of Solids, vol. 11, pp. 127–140.

39. V. Tvergaard (1982) "On localization in ductile materials containing spherical voids," International Journal of Fracture, vol. 18, pp. 237–252.

40. T. Christman, A. Needleman and S. Suresh (1989) "An experimental and numerical study of deformation in metal-ceramic composites," Acta Metallurgica et Materialia, vol. 37, pp. 3029–3050.

41. V. Tvergaard (1990) "Analysis of tensile properties for a whisker-reinforced metal matrix composites," Acta Metallurgica et Materialia, vol. 38, pp. 185–194.

42. G. Bao, J. W. Hutchinson and R. M. McMeeking (1991) "Particle reinforcement of ductile matrices against plastic flow and creep," Acta Metallurgica et Materialia, vol. 39, pp. 1871–1882.

43. G. L. Povirk, A. Needleman and S. R. Nutt (1991) "An analysis of the effect of residual stresses on deformation and damage mechanisms in Al-SiC composites," Materials Science and Engineering A, vol. 132, pp. 31–38.

44. J. Llorca, A. Needleman and S. Suresh (1991) "An analysis of the effects of matrix void growth on deformation and ductility in metal-ceramic composites," Acta Metallurgica et Materialia, vol. 39, pp. 2317–2335.

45. M. B. Bush (1992) "An investigation of two- and three-dimensional models for predicting the elastic properties of particulate- and whisker-reinforced composite materials," Materials Science and Engineering A, vol. 154, pp. 139–148.

46. Y.-L. Shen, M. Finot, A. Needleman and S. Suresh (1994) "Effective elastic response of two-phase composites," Acta Metallurgica et Materialia, vol. 42, pp. 77–97.

47. M. Taya and R. J. Arsenault (1989) Metal matrix composites – thermomechanical behavior, Pergamon Press, New York.

48. S. Suresh, A. Mortensen and A. Needleman (1993) Fundamentals of metal matrix composites, Butterworth-Heinemann, Stoneham, MA.

49. T. W. Clyne and P. J. Withers (1993) An introduction to metal matrix composites, Cambridge University Press, Cambridge.

50. N. Chawla and Y.-L. Shen (2001) "Mechanical behavior of particle reinforced metal matrix composites," Advanced Engineering Materials, vol. 3, pp. 357–370.

51. D. C. Drucker (1966) "The continuum theory of plasticity on the macroscale and the microscale," Journal of Materials, vol. 1, pp. 873–910.

52. D. C. Drucker (1965) "Engineering and continuum aspects of high strength materials," in V. F. Zackay: High Strength Materials, pp. 795–833, Wiley, New York.

53. T. W. Butler and D. C. Drucker (1973) "Yield strength and microstructural scale: A continuum study of pearlitic versus spheroidized steel," Journal of Applied Mechanics, vol. 40, pp. 780–784.

54. Y.-L. Shen, M. Finot, A. Needleman and S. Suresh (1995) "Effective plastic response of two-phase composites," Acta Metallurgica et Materialia, vol. 43, pp. 1701–1722.

55. C. L. Hom (1992) "Three-dimensional finite element analysis of plastic deformation in a whisker-reinforced metal matrix composite," Journal of the Mechanics and Physics of Solids, vol. 40, pp. 991–1008.

56. A. Levy and J. M. Papazian (1990) "Tensile properties of short fiber-reinforced SiC/Al composites: Part II. Finite element analysis," Metallurgical Transactions A, vol. 21A, pp. 411–420.

57. E. Weissenbek, H. J. Bohm and F. G. Rammerstorfer (1994) "Micromechanical investigations of arrangement effects in particle reinforced metal matrix composites," Computational Materials Science, vol. 3, pp. 263–278.

58. J. Segurado, C. Gonzalez and J. Llorca (2003) "A numerical investigation of the effect of particle clustering on the mechanical properties of composites," Acta Materialia, vol. 51, pp. 2355–2369.

59. N. Chawla, R. S. Sidhu and V. V. Ganesh (2006) "Three-dimensional visualization and microstructure-based modeling of deformation in particle reinforced composites," Acta Materialia, vol. 54, pp. 1541–1548.

60. H. P. Ganser, F. D. Fischer and E. A. Werner (1998) "Large strain behavior of two-phase materials with random inclusions," Computational Materials Science, vol. 11, pp. 221–226.

61. D. B. Zahl and R. M. McMeeking (1991) "The influence of residual stress on the yielding of metal matrix composites," Acta Metallurgica et Materialia, vol. 39, pp. 1117–1122.

62. A. Levy and J. M. Papazian (1991) "Elastoplastic finite element analysis of short-fiber reinforced SiC/Al composites – effects of thermal treatment," Acta Metallurgica et Materialia, vol. 39, pp. 2255–2266.

63. I. Dutta, J. D. Sims and D. M. Seigenthaler (1993) "An analytical study of residual-stress effects on uniaxial deformation of whisker reinforced metal-matrix composites," Acta Metallurgica et Materialia, vol. 41, pp. 885–908.

64. N. Shi, B. Wilner and R. J. Arsenault (1992) "An FEM study of the plastic deformation processes of whisker reinforced SiC/Al composites," Acta Metallurgica et Materialia, vol. 40, pp. 2841–2854.

65. L. C. Davis and J. E. Allison (1993) "Residual stresses and their effects on deformation in particle-reinforced metal-matrix composites," Metallurgical Transactions A, vol. 24A, pp. 2487–2496.

66. N. Chawla, U. Habel, Y.-L. Shen, C. Andres, J. W. Jones and J. E. Allison (2000) "The effect of matrix microstructure on the tensile and fatigue behavior of SiC particle-reinforced 2080 Al matrix composites," Metallurgical and Materials Transactions A, vol. 31A, pp. 531–540.

67. N. Chawla, C. Andres, L. C. Davis, J. W. Jones and J. E. Allison (2000) "The interactive role of inclusions and SiC reinforcement on the high-cycle fatigue resistance of particle reinforced metal matrix composites," Metallurgical and Materials Transactions A, vol. 31A, pp. 951–957.

68. T. Wilkins and Y.-L. Shen (2001) "Stress enhancement at inclusion particles in aluminum matrix composites: computational modeling and implications to fatigue damage," Computational Materials Science, vol. 22, pp. 291–299.

69. V. M. Levin (1967) "On the coefficient of thermal expansion of heterogeneous materials," Mechanics of Solids, vol. 2, pp. 58–61.

70. R. A. Schapery (1968) "Thermal expansion coefficients of composite materials based on energy principles," Journal of Composite Materials, vol. 2, pp. 380–404.

71. B. W. Rosen and Z. Hashin (1970) "Thermal expansion coefficients and specific heats of composite materials," International Journal of Engineering Sciences, vol. 8, pp. 157–173.

72. E. H. Kerner (1956) "The elastic and thermo-elastic properties of composite media," Proceedings of the Physical Society B, vol. 69, pp. 808–813.

73. Y.-L. Shen, A Needleman and S. Suresh (1994) "Coefficients of thermal expansion of metal-matrix composites for electronic packaging," Metallurgical and Materials Transactions A, vol. 25A, pp. 839–850.

74. C. Zweben (1992) "Metal-matrix composites for electronic packaging," JOM, vol. 44, pp. 15–23.

75. M. K. Premkumar, W. H. Hunt and R. R. Sawtell (1992) "Aluminum composite – materials for multichip modules," JOM, vol. 44, pp. 24–48.

76. C. Zweben (2005) "Electronic packaging materials," Advanced Materials and Processes, vol. 163, No. 10, pp. 33–37.

77. Y.-L. Shen (1998) "Thermal expansion of metal–ceramic composites: a three-dimensional analysis," Materials Science and Engineering A, vol. 252, pp. 269–275.

78. M. Olsson, A. E. Giannakopoulos and S. Suresh (1995) "Elastoplastic analysis of thermal cycling: Ceramic particles in a metallic matrix," Journal of the Mechanics and Physics of Solids, vol. 43, pp. 1639–1671.

79. D. K. Balch, T. J. Fitzgerald, V. J. Michaud, A. Mortensen, Y.-L. Shen and S. Suresh (1996) "Thermal expansion of metals reinforced with ceramic particles and microcellular foams," Metallurgical and Materials Transactions A, vol. 27A, pp. 3700–3717.

80. Y.-L. Shen (1997) "Combined effects of microvoids and phase contiguity on the thermal expansion of metal-ceramic composites," Materials Science and Engineering A, vol. 237, pp. 102–108.

81. J. H. Lupinski and R. S. Moore (1989) Polymeric materials for electronics packaging and interconnection, American Chemical Society, Washington.

82. M. G. Pecht, L. T. Nguyen and E. B. Hakim (1995) Plastic-encapsulated microelectronics, Wiley, New York.

83. M. Chaturvedi and Y.-L. Shen (1998) "Thermal expansion of particle-filled plastic encapsulant: A micromechanical characterization," Acta Materialia, vol. 46, pp. 4287–4302.

84. W. M. Wolverton (1987) "The mechanisms and kinetics of solder joint degradation," Brazing and Soldering, vol. 13, pp. 33–38.

85. D. Tribula, D. Grivas, D. R. Frear and J. W. Morris, Jr. (1989) "Microstructural observations of thermomechanically deformed solder joints," Welding Research Supplement, October, pp. 404s–409s.

86. J. H. Lau (1991) Solder joint reliability: theory and applications, Van Nostrand Reinhold, New York.

87. D. R. Frear, H. Morgan, S. Burchett and J. Lau (1994) The mechanics of solder alloy interconnects, Van Nostrand Reinhold, New York.

88. W. J. Plumbridge (1996) "Solders in electronics," Journal of Materials Science, vol. 119, pp. 2501–2514.

89. X. W. Liu and W. J. Plumbridge (2003) "Damage produced in model solder (Sn-37Pb) joints during thermomechanical cycling," Journal of Electronic Materials, vol. 32, pp. 278–286.

90. W. D. Callister, Jr. (2006) Materials science and engineering: an introduction, 7th ed., Wiley, New York.

91. R. Abbaschian, L. Abbaschian and R. E. Reed-Hill (2009) Physical metallurgy principles, 4th ed., Cengage Learning, Stamford, CT.

92. Y.-L. Shen, W. Li and H. E. Fang (2001) "Phase structure and cyclic deformation in eutectic tin-lead alloy: A numerical analysis," Journal of Electronic Packaging, vol. 123, pp. 74–78.

93. M. A. Dudek, R. S. Sidhu, N. Chawla and M. Renavikar (2006) "Microstructure and mechanical behavior of novel rare earth-containing Pb-free solders," Journal of Electronic Materials, vol. 35, pp. 2088–2097.

94. A. Fisher-Cripps (2002) Nanoindentation, Springer, New York.

95. W. C. Oliver and G. M. Pharr (1992) "An improved technique for determining hardness and elastic modulus using load and displacement sensing indentation experiments," Journal of Materials Research, vol. 7, pp. 1564–1538.

96. W. C. Oliver and G. M. Pharr (2004) "Measurement of hardness and elastic modulus by instrumented indentation: Advances in understanding and refinements to methodology," Journal of Materials Research, vol. 19, pp. 3–20.

97. C. A. Schuh (2006) "Nanoindentation studies of materials," Materials Today, vol. 9(5), pp. 32–40.

98. A. Gouldstone, N. Challacoop, M. Dao, J. Ki, A. M. Minor and Y.-L. Shen (2007) "Indentation across size scales and disciplines: recent developments in experimentation and modeling," Acta Materialia, vol. 55, pp. 4015–4039.

99. A. Misra, M. Verdier, Y. C. Lu, H. Kung, T. E. Mitchell, M. Nastasi and J. D. Embury (1998) "Structure and mechanical properties of Cu-X (X = Nb, Cr, Ni) nanolayered composites," Scripta Materialia, vol. 39, pp. 555–560.

100. A. Misra and H. Kung (2001) "Deformation behavior of nanostructured metallic multilayers," Advanced Engineering Materials, vol. 3, pp. 217–222.

101. A. Misra, M. J. Demkowicz, J. Wang and R. G. Hoagland (2008) "The multiscale modeling of plastic deformation in metallic nanolayered composites," JOM, vol. 60(4), pp. 39–42.

102. X. H. Tan and Y.-L. Shen (2005) "Modeling analysis of the indentation-derived yield properties of metallic multilayered composites," Composites Science and Technology, vol. 65, pp. 1639–1646.

103. G. Tang, Y.-L. Shen, D. R. P. Singh and N. Chawla (2010) "Indentation behavior of metal-ceramic multilayers at the nanoscale: Numerical analysis and experimental verification," Acta Materialia, doi: 10.1016/j.actamat.2009.11.046.

104. Y.-L. Shen and Y. L. Guo (2001) "Indentation modelling of heterogeneous materials," Modelling and Simulation in Materials Science and Engineering, vol. 9, pp. 391–398.

105. B. D. Kozola and Y.-L Shen (2003) "A mechanistic analysis of the correlation between overall strength and indentation hardness in discontinuously reinforced aluminum," Journal of Materials Science, vol. 38, pp. 901–907.

106. Y.-L. Shen, J. J. Williams, G. Piotrowski, N. Chawla and Y. L. Guo (2001) "Correlation between tensile and indentation behavior of particle-reinforced metal matrix composites: an experimental and numerical study," Acta Materialia, vol. 49, pp. 3219–3229.

107. R. Pereyra and Y.-L. Shen (2004) "Characterization of particle concentration in indentation-deformed metal-ceramic composites," Materials Characterization, vol. 53, pp. 373–380.

Chapter 7
Challenges and Outlook

Constrained deformation of materials is a ubiquitous phenomenon in the engineering world. In a majority of situations it creates real problems that limit the performance of materials and devices and generate reliability concerns; in others the problem is actually induced by design for the purpose of achieving specific functionalities. In previous chapters we have treated thermo-mechanical deformation influenced by physical constraint due to the outside media bonded to the material or the internal material heterogeneities. Attempts were made to establish a unified theme encompassing past developments as well as new analyses. The deformation characteristics were addressed from a continuum modeling point of view, without elaboration on the attributes of microstructural details (molecular configuration, crystal structure, grain size, texture, crystal defect density, defect interaction etc.). Although this is a simplified way to deal with the overwhelmingly complex problem spanning a wide range of length scales, it serves the critical purpose of facilitating a mechanistic framework for basic understanding, which is frequently lacking even in the research community. It goes without saying that, in real-life engineering design and analysis, only the simple approach matters.

There are, however, fundamental issues specific to the modeling of constrained deformation which warrant additional considerations. Some of these are highlighted in the sections below.

7.1 Material Parameters and Constitutive Models

The output of a model can only be as good as its input allows. When uncertainty exists in the model setup, the risk of having problematic results from the modeling inevitably rises. This can become an intricate issue when dealing with miniaturized structures and devices. Some mechanical properties of materials are a function of their physical size. A prominent example is the plastic yielding properties of metallic materials. The plastic flow stress is a function of specimen size and the microstructure feature size [1–3], so care must be taken in "assigning" the yield strength value in the analysis. The elastic modulus, on the other hand, is typically independent of the

Y.-L. Shen, *Constrained Deformation of Materials: Devices, Heterogeneous Structures and Thermo-Mechanical Modeling*, DOI 10.1007/978-1-4419-6312-3_7,
© Springer Science+Business Media, LLC 2010

specimen size and microstructure. However, for materials in their thin-film form, even the elastic modulus is subject to uncertainty due to the possible processing-induced defects after deposition [4, 5]. In certain cases the importance of including strain hardening with an appropriate plasticity model, as discussed in Chap. 3, also needs to be emphasized. Furthermore, the *true* plastic response under constrained cyclic thermal and/or mechanical conditions remains largely unexplored, let alone understood. Additional complications may arise when the time-dependent response needs to be accounted for, especially for metals at elevated temperatures and for polymers in general.

Even with a fixed size and shape, a material may have a different mechanical behavior in the bonded form, compared to its free-standing counterpart. This concern applies to the externally bonded material (such as in thin-film devices) and materials with a heterogeneous internal structure (such as composite materials). Our current understanding of the interface characteristics is very limited. For instance, how plastic flow really takes place near the interface, and how the external constraint (due to adjacent thick, thin, stiff, or compliant materials) is transmitted through the interface are in need of continued experimental and numerical explorations. Atomic scale modeling is apparently important in such an endeavor. However a fundamental question will remain: under what circumstances do continuum-level analyses require interface constitutive laws which warrant quantitative relations obtained from the atomistic scale? The linkage of phenomena pertaining to disparate length scales is still vague. Since continuum-level modeling will remain to be a dominant engineering tool, its capability of handling localized interfacial behavior at very small length scales will continue to be refined and assessed [6].

To this end, it is worth mentioning that various theoretical and computational treatments at above the atomistic scale – including those on the basis of discrete dislocation dynamics and continuum-level strain gradient plasticity theory – have been developed (see, e.g., [7–34]). Some of these studies were also aimed at investigating the size effect, which can be a consequence of constrained plastic deformation mediated by the interface. The strain gradient plasticity theory in the present context is predicated upon the notion that a well bonded interface leads to the buildup of dislocations in the metallic material and thus a plastic strain gradient. The plastic strain gradient enhances the resistance to continued plastic flow, the quantification of which requires additional characteristic length parameters.

7.2 Technological Relevance

Almost all the topics and discussion points included in this book bear significance in modern technology. Indeed advanced technology has been, and will continue to be, the driving force of fundamental research, for meeting the challenges described in the previous section. A case in point is thin-film metallization in microelectronics (cf. Chap. 4). Over the last several decades, the field of microelectronic interconnects has offered tremendous technologically relevant opportunities for researches in thin films and small-scale mechanical behavior to prosper. In turn the microelectronics industry has benefited from the expanding knowledge base for improved design and material selection/development. The various modeling examples presented in previous chapters have demonstrated their capability of providing useful insight on physical features and experimental observations in both the industrial and laboratory settings.

One can only expect that the continued advancement of micro- and nano-scale devices and material systems will rely even more on our understanding and control of constrained deformation, to avoid its detrimental consequences and/or exploit its potential advantages. Developing computational and experimental tools to address these issues will certainly provide challenging opportunities for future endeavor.

7.3   Learning from Nature

While the focus of this book is on engineering applications, it helps to mention that constrained deformation is omnipresent in nature. Most natural or biological materials possess complex heterogeneous structures. They often display superior mechanical properties and failure resistance, although their constituents may be weak and fragile as a standalone macroscopic material. The structural hierarchy, mechanical characteristics, multiple functionalities, and self-adapting abilities of natural materials have begun to inspire researchers in the quest for designing novel or biomimetic materials. At the heart of this venture is the correlation between constrained deformation and multi-scale material architecture [35–45].

A characteristic example is bone, which is basically a ceramic–polymer composite. The ceramic is a mineral phase of hydroxyapatite, and the polymer is collagen protein. However, their structure is hierarchical with extreme complexity, and is only partially understood. The fundamental structure of the soft organic component is the collagen molecules, which provide the flexibility and energy dissipation capability during deformation. Staggered arrays of collagen molecules form collagen fibrils, which are intercalated with small crystals of hydroxyapatite mineral. The mineralized collagen fibrils are themselves arranged into arrays, joined by the protein phase as the matrix to form fibers. The fibers pattern into bundles which are in turn organized into a cylindrical lamellar structure termed osteon. The osteons may be viewed as the building block of macroscopic bone.

If one focuses on the individual fibrils and their array, the structure can then be depicted as in Fig. 7.1. The left part is the array of mineralized collagen fibrils, with the dark regions representing the intrafibrillar and extrafibrillar mineral platelets. The right part of the figure is a zoom-in of a fibril, showing the familiar particle-matrix composite structure. The thickness of each mineral platelet is typically 1–4 nm with as aspect ratio of about 30–40. The volume fraction of the mineral phase is typically around 40–50%. It is immediately evident that, under external loading, the constrained elastic and inelastic deformation may be analyzed as a two-phase composite, using the approach discussed extensively in Chap. 6. Indeed

Collagen matrix

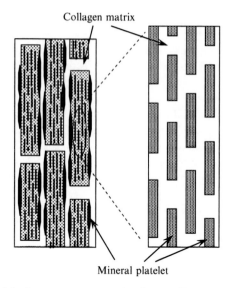

Mineral platelet

Fig. 7.1 Schematic of the bone nanostructure, showing the fibril arrays (*left*) and the mineral crystal platelets/collagen matrix structure inside a fibril (*right*). The mineral platelet has a thickness of about 1–4 nm and an aspect ratio of about 30–40. The diameter of a fibril is on the order of 100 nm [35, 44]

simplified 2D analyses on the effective elastic properties of the mineral-collagen nano-composite structure in bone have been reported [46, 47]. Seeking a deeper understanding would require more sophisticated numerical models, possibly 3D, and innovative ways to tackle the hierarchy inherent in the bone structure.

It is anticipated that constrained deformation in biological materials will continue to attract increasing attention on all of the experimental, theoretical and computational fronts. Virtually an unlimited source of insights can be cultivated by studying the mechanical behavior and underpinning architecture in natural systems. The diverse knowledge base thus gained may form the basis for future innovation of high-performance engineering materials and devices.

7.4 Beyond Deformation

Throughout this book attention is directed toward deformation characteristics of heterogeneous structures and materials. Information on deformation provides a simple connection to the tendency of failure (damage) initiation, which is of great concern under most circumstances. However, direct simulation of material failure would offer a straightforward picture without ambiguity, contributing to the true predictive capability of a model. Even before final failure, any local damage inside the structure would alter the stress and deformation fields, rendering a pure deformation-based analysis inadequate.

Numerical modeling of material failure at the continuum level is a rapidly evolving field. While constitutive models and their implementation techniques have been developed for various types of fracture (e.g., interfacial decohesion, ductile damage etc.), their association with realistic material behavior is still rather weak. Incorporating explicit failure features in modeling no doubt makes excellent research tools (see, e.g., [48–54] for fracture evolution in particle-reinforced composites), but it is still far from becoming a routine practice in engineering analysis. One of the primary issues is the fidelity of material parameters needed as input for the damage models. We end our discussion by presenting the following final case study as an illustration.

7.4.1 Case Study: Ductile Failure in Solder Joint

7.4.1.1 Model Setup

Deformation analyses of solder joints, in particular the lap-shear test configuration, were discussed extensively in Chap. 5. Here a ductile damage feature is added to the model, which, in conjunction with the element removal function in the explicit finite element scheme, is capable of simulating direct failure of the solder. Figure 7.2a shows a schematic of the model geometry, with the solder material bonded to two copper (Cu) substrates. The width (w) and thickness (t) of the solder joint are 1 and 0.5 mm, respectively, and the substrate dimensions H and W are 2.5 and 0.5 mm, respectively. At each interface between the solder and Cu substrate, there is a 5-μm thick intermetallic layer included in the model (not specifically shown in Fig. 7.2a). Monotonic deformation of solder is induced by prescribing a constant boundary velocity, v_x, at the far right edge of the lower substrate. This will give rise to a nominal shear deformation in solder. The nominal shear strain rate imposed on the solder joint is thus $\frac{v_x}{t}$. During deformation the x-direction movement of the far left edge of the upper Cu substrate is forbidden, but movement in the y-direction is allowed except that the upper-left corner of the upper substrate is entirely fixed. The top boundary of the upper Cu is also fixed in the y-direction but its x-movement is allowed. The plane strain deformation condition is assumed.

In the model the Cu substrate is taken to be elastic, with Young's modulus of 114 GPa, Poisson's ratio of 0.31 and density of 8,930 kg/m³. The Cu_6Sn_5 intermetallic layer is also assumed to be elastic, with Young's modulus of 85.5 GPa, Poisson's ratio of 0.28 and density of 8,280 kg/m³. The solder, taken to be the Sn-1.0Ag-0.1Cu alloy as in Sect. 5.2.2, is elastic-viscoplastic with Young's modulus of 47 GPa, Poisson's ratio of 0.36 and density of 5,760 kg/m³. Its yielding and strain hardening response follows the experimental stress–strain curves for different strain rates [55]. At or below the strain rate of 0.005 s⁻¹, the initial yield strength is 20 MPa; the flow strength increases to a peak value of 36 MPa at the plastic strain of 0.15, beyond which a perfectly plastic behavior is assumed. This slow-rate form

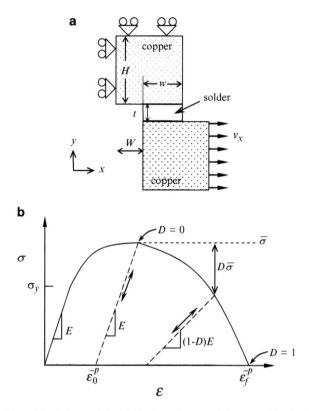

Fig. 7.2 (a) The solder joint model and the boundary conditions used in the finite element analysis. (b) Schematic showing the ductile damage response, in terms of the uniaxial stress–strain curve

is considered as the "static" response. The rate-dependent plastic flow stress follows (2.20), rewritten here as

$$\sigma_e = h(\bar{\varepsilon}^p) \cdot R(\frac{d\bar{\varepsilon}^p}{dt}), \tag{7.1}$$

where σ_e is the von Mises effective stress, h (a function of equivalent plastic strain $\bar{\varepsilon}^p$) is the static plastic stress–strain response, and R (a function of plastic strain rate $\frac{d\bar{\varepsilon}^p}{dt}$) defines the ratio of flow stress at higher strain rates to the static flow stress where R equals unity. Compared to rate-independent plasticity, this formulation utilizes the scaling parameter R to quantify the "strain rate hardening" effect. In the present case R values are 1.0, 1.9, 2.4, 2.8, 3.1, 3.4 and 3.5 at the plastic strain rates of, respectively, 0.005, 0.5, 6, 50, 100, 200 and 300 s^{-1}.

A progressive ductile damage model is utilized to simulate failure of the solder alloy [56, 57]. Figure 7.2b shows a schematic of the stress–strain curve which includes the damage response (solid curve). The damage process is quantified by a scalar damage parameter D, with

$$\sigma = (1 - D)\bar{\sigma}, \tag{7.2}$$

where σ is the current flow stress and $\bar{\sigma}$ is the flow stress in the absence of damage. In addition to leading to softening of the plastic stress, damage is also manifested by the degradation of elastic modulus as shown by the dashed unloading/reloading line in Fig. 7.2b. The equivalent plastic strains are $\bar{\varepsilon}_0^p$ and $\bar{\varepsilon}_f^p$ at the onset of damage ($D=0$) and failure ($D=1$), respectively. A material element loses its stress-carrying capability when its D attains unity, at which point the element will be removed from the mesh so a tiny void thus develops. Cracking is then a consequence of linking multiple adjacent voids in the model.

   Upon damage initiation, strain softening and thus strain localization set in, which results in a strong mesh dependency. To alleviate the problem, a characteristic length L is used in the model, with

$$\bar{u}^p = L\bar{\varepsilon}^p \tag{7.3}$$

where $\bar{u}^p$ represents a plastic displacement quantity, and L is defined as the square root of the integration point area in each finite element. The softening phenomenon is now expressed as a stress–displacement relationship [58]. Prior to the initiation of damage, $\bar{u}^p = 0$; after damage initiation (7.3) starts to take effect. Removal of the element (failure) occurs when $\bar{u}^p$ reaches the specified failure value, $\bar{u}_f^p$. The evolution of the damage parameter is taken to follow a linear form,

$$D = \frac{\bar{u}^p}{\bar{u}_f^p}. \tag{7.4}$$

The damage response is thus completely specified by the two parameters $\bar{\varepsilon}_0^p$ and $\bar{u}_f^p$. In the present illustration they are chosen to be 0.18 and 3 µm, respectively. It is noted that the chosen value of $\bar{u}_f^p$ corresponds to a $\bar{\varepsilon}_f^p$ value of approximately 0.5. These parameters are based on some measured tensile stress–strain curves of bulk pure Sn or Sn-rich solder alloy.

7.4.1.2 Plastic Deformation Field

We now present the modeling result. Figure 7.3a and b show the contour plots of equivalent plastic strain along with the cracks (removed elements), during the cracking process and upon final failure, respectively, for the 10 s^{-1} applied shear strain rate. The nominal shear strains corresponding to parts (a) and (b) are 0.275 and 0.295, respectively. Note that final failure occurs when a major crack traversing the entire span of the solider joint is formed. The initiation of cracks appears at the four corners of the solder joint. Crack propagation follows the path of greatest equivalent plastic strain, which in turn is evolving during the deformation and

a

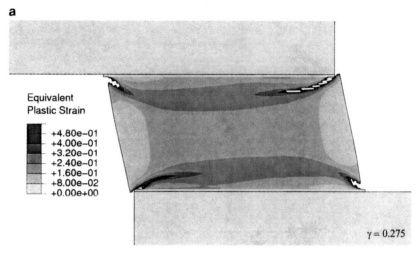

b

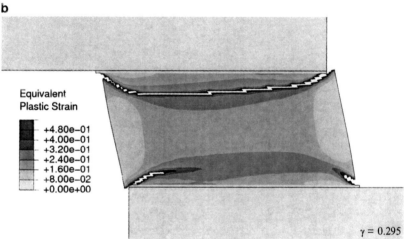

Fig. 7.3 Contour plots of equivalent plastic strain and the evolution of failure, (**a**) during the cracking process (at the nominal shear strain 0.275) and (**b**) upon final failure (at the nominal shear strain 0.295)

damage processes. It is seen that the cracks first tend to grow inward along the near 45° direction. However, due to the dominant shear mode, a band of strong plasticity in the horizontal direction gradually forms parallel to each interface. Damage therefore localizes along the band where final linkage of cracks takes place.

It is worth having a comparison between the results with and without the failure model included in the simulation. Figure 7.4 shows the contour plot of equivalent plastic strain in the solder joint at the nominal shear strain of 0.295, under the same 10 s^{-1} applied shear strain rate, with the damage portion of the model "switched off." Note that the nominal shear strain is the same as in Fig. 7.3b. Without the progressive damage and fracture, the plastic strain field becomes more uniform

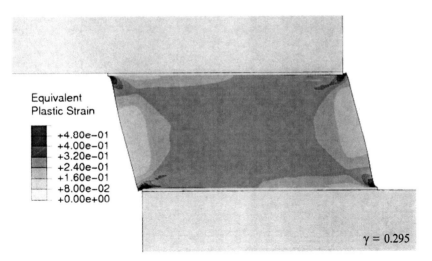

Fig. 7.4 Contour plots of equivalent plastic strain in solder at the nominal shear strain of 0.295, if no damage feature is included in the finite element model

(less localized) in Fig. 7.4. Under the current high strain rate, the strain-rate hardening effect is more pronounced and is not counteracted by local fracture which promotes strain concentration. The banded structure is thus no longer apparent.

We point out that the result in Fig. 7.3 is based on the ductile damage model only; the model does not include other failure features such as debonding along the interface. This more brittle form of damage may be added to the simulation to address more realistic situations. Even with the ductile damage model utilized here, there are major concerns regarding the choice of material parameters. In general $\overline{\varepsilon}_0^p$ is a function of stress triaxiality, σ_H/σ_e, where $\sigma_H = \left(\sigma_{xx} + \sigma_{yy} + \sigma_{zz}\right)/3$ is the hydrostatic stress and σ_e is the von Mises effective stress. In the above analysis $\overline{\varepsilon}_0^p$ is assumed to be independent of stress triaxiality, simply because of the lack of experimental data that may be used for defining the functional form. Furthermore, both the parameters $\overline{\varepsilon}_0^p$ and $\overline{u}_f^p$ for the damage model are chosen on a rather crude basis of tensile stress–strain curves of bulk specimens. It is understood that bulk materials and actual solder joints have significantly different physical sizes and microstructures and thus different constitutive responses. In addition, tensile loading tends to promote microvoid nucleation and coalescence, resulting in easier damage compared to other forms of loading. The linear form of (7.4) is another possible oversimplification.

Hence, with a capable failure model at hand, the simulation still has to be settled on somewhat arbitrary material parameters. Although the cracking morphology in Fig. 7.3 is quite similar to many experimental observations (an example for cyclic shearing in shown in Fig. 5.15), the uncertain set of input parameters does cast doubt on the true capability of the numerical model for quantitative life prediction of engineering components. Developing a better connection between theory and useful material database, for analyzing constrained deformation leading to failure, certainly offers challenging opportunities for fundamental research as well as engineering applications.

References

1. W. D. Nix (1989) "Mechanical properties of thin films," Metallurgical Transactions A, vol. 20A, pp. 2217–2245.
2. E. Arzt (1998) "Size effects in materials due to microstructural and dimensional constraints: a comparative review," Acta Materialia, vol. 46, pp. 5611–5626.
3. L. B. Freund and S. Suresh (2003) Thin film materials – stress, defect formation and surface evolution, Cambridge University Press, Cambridge.
4. H. B. Huang and F. Spaepen (2000) "Tensile testing of free-standing Cu, Ag and Al thin films and Ag/Cu multilayers," Acta Materialia, vol. 48, pp. 3261–3269.
5. D. T. Read (1998) "Young's modulus of thin films by speckle interferometry," Measurement Science and Technology, vol. 9, pp. 676–685.
6. Y.-L. Shen (2008) "Externally constrained plastic flow in miniaturized metallic structures: A continuum-based approach to thin films, lines and joints," Progress in Materials Science, vol. 53, pp. 838–891.
7. J. Hurtado and L. B. Freund (1999) "The force on a dislocation near a weakly bonded interface," Journal of Elasticity, vol. 52, pp. 167–180.
8. J. Y. Shu, N. A. Fleck, E. Van der Giessen and A. Needleman (2001) "Boundary layers in constrained plastic flow: comparison of nonlocal and discrete dislocation plasticity," Journal of the Mechanics and Physics of Solids, vol. 49, pp. 1361–1395.
9. B. von Blanckenhagen, P. Gumbsch and E. Arzt (2001) "Dislocation sources in discrete dislocation simulations of thin-film plasticity and the Hall-Petch relation," Modelling and Simulation in Materials Science and Engineering, vol. 9, pp. 157–169.
10. V. Weihnacht and W. Bruckner (2001) "Dislocation accumulation and strengthening in Cu thin films," Acta Materialia, vol. 49, pp. 2365–2372.
11. L. P. Kubin and A. Mortensen (2003) "Geometrically necessary dislocations and strain-gradient plasticity: a few critical issues," Scripta Materialia, vol. 48, pp. 119–125.
12. L. Nicola, E. Van der Giessen and A. Needleman (2003) "Discrete dislocation analysis of size effects in thin films," Journal of Applied Physics, vol. 93, pp. 5920–5928.
13. P. Pant, K. W. Schwarz and S. P. Baker (2003) "Dislocation interactions in thin FCC metal films," Acta Materialia, vol. 51, pp. 3243–3258.
14. P. Gudmundson (2004) "A unified treatment of strain gradient plasticity," Journal of the Mechanics and Physics of Solids, vol. 52, pp. 1379–1406.
15. M. J. Buehler, A. Hartmaier and H. Gao (2004) "Hierarchical multi-scale modeling of plasticity of submicron thin metal films," Modelling and Simulation in Materials Science and Engineering, vol. 12, pp. S391–S413.
16. H. D. Espinosa, S. Berbenni, M. Panico and K. W. Schwarz (2005) "An interpretation of size-scale plasticity in geometrically confined systems," Proceedings of the National Academy of Sciences of the United States of America, vol. 47, pp. 16933–16938.
17. P. Fredriksson and P. Gudmundson (2005) "Size-dependent yield strength of thin films," International Journal of Plasticity, vol. 21, pp. 1834–1854.
18. S. Yefimov and E. Van der Giessen (2005) "Size effects in single crystal thin films: nonlocal crystal plasticity simulations," European Journal of Mechanics A – Solids, vol. 24, pp. 183–193.
19. X. Han and N. M. Ghoniem (2005) "Stress field and interaction forces of dislocations in anisotropic multilayer thin films," Philosophical Magazine, vol. 85, pp. 1205–1225.
20. N. M. Ghoniem and X. Han (2005) "Dislocation motion in anisotropic multilayer materials," Philosophical Magazine, vol. 85, pp. 2809–2830.
21. G. Yun, K. C. Hwang, Y. Huang, P. D. Wu and C. Liu (2006) "Size effect in tension of thin films on substrate: a study based on the reformulation of mechanism-based strain gradient plasticity," Philosophical Magazine, vol. 86, pp. 5553–5566.
22. Y. Xiang and J. J. Vlassak (2006) "Bauschinger and size effects in thin-film plasticity," Acta Materialia, vol. 54, pp. 5449–5460.

23. L. Nicola, Y. Xiang, J. J. Vlassak, E. Van der Giessen and A. Needleman (2006) "Plastic deformation of freestanding thin films: experiments and modeling," Journal of the Mechanics and Physics of Solids, vol. 54, pp. 2089–2110.

24. Y. Shen and P. M. Anderson (2006) "Transmission of a screw dislocation across a coherent, slipping interface," Acta Materialia, vol. 54, pp. 3941–3951.

25. Y. Shen and P. M. Anderson (2007) "Transmission of a screw dislocation across a coherent, non-slipping interface," Journal of the Mechanics and Physics of Solids, vol. 55, pp. 956–979.

26. R. K. Abu Al-Rub, G. Z. Voyiadjis and D. J. Bammann (2007) "A thermodynamic based higher order gradient theory for size dependent plasticity," International Journal of Solids and Structures, vol. 44, pp. 2888–2923.

27. J. Douin, F. Pettinari-Sturmel and A. Coujou (2007) "Dissociated dislocations in confined plasticity," Acta Materialia, vol. 55, pp. 6453–6458.

28. F. Akasheh, H. M. Zbib, J. P. Hirth, R. G. Hoagland and A. Misra (2007) "Interactions between glide dislocations and parallel interfacial dislocations in nanoscale strained layers," Journal of Applied Physics, vol. 102, 034314.

29. R. K. Abu Al-Rub (2008) "Modeling the interfacial effect on the yield strength and flow stress of thin metal films on substrates," Mechanics Research Communications, vol. 35, pp. 65–72.

30. M. Kuroda and V. Tvergaard (2008) "On the formulation of higher-order strain gradient crystal plasticity models," Journal of the Mechanics and Physics of Solids, vol. 56, pp. 1591–1608.

31. P. Fredriksson and P. L. Larsson (2008) "Wedge indentation of thin films modelled by strain gradient plasticity," International Journal of Solids and Structures, vol. 45, pp. 5556–5566.

32. D. Weygand, M. Poignant, P. Gumbsch and O. Kraft (2008) "Three-dimensional dislocation dynamics simulation of the influence of sample size on the stress-strain behavior of fcc single-crystalline pillars," Materials Science and Engineering A, vol. 483, pp. 188–190.

33. A. G. Evans and J. W. Hutchinson (2009) "A critical assessment of theories of strain gradient plasticity," Acta Materialia, vol. 57, pp. 1675–1688.

34. Z. H. Li, C. T. Hou, M. S. Huang and C. J. Ouyang (2009) "Strengthening mechanism in micro-polycrystals with penetrable grain boundaries by discrete dislocation dynamics simulation and Hall-Petch effect," Computational Materials Science, vol. 46, pp. 1124–1134.

35. P. Fratzl and R. Weinkamer (2007) "Nature's hierarchical materials," Progress in Materials Science, vol. 52, pp. 1263–1334.

36. M. A. Meyers, P.-Y. Chen, A. Y.-M. Lin and Y. Seki (2008) "Biological materials: Structure and mechanical properties," Progress in Materials Science, vol. 53, pp. 1–206.

37. A. V. Srinivasan, G. K. Haritos and F. L. Hedberg (1991) "Biomimetics: Advancing man-made materials through guidance from nature," Applied Mechanics Review, vol. 44, pp. 463–482.

38. U. G. K. Wegst and M. F. Ashby (2004) "The mechanical efficiency of natural materials," Philosophical Magazine, vol. 84, pp. 2167–2181.

39. H. Gao, B. Ji, I. L. Jager, E. Arzt and P. Fratzl (2003) "Materials become insensitive to flaws at nanoscale: Lessons from nature," Proceedings of the National Academy of Sciences of the United States of America, vol. 100, pp. 5597–5600.

40. G. Mayer (2005) "Rigid biological systems as models for synthetic composites," Science, vol. 310, pp. 1144–1147.

41. K. Tai, M. Dao, S. Suresh, A. Palazoglu and C. Ortiz (2007) "Nanoscale heterogeneity promotes energy dissipation in bone," Nature Materials, vol. 6, pp. 454–462.

42. C. Ortiz and M. C. Boyce (2008) "Bioinspired structural materials," Science, vol. 319, pp. 1053–1054.

43. M. E. Launey, E. Munch, D. H. Alsem, H. B. Barth, E. Saiz, A. P. Tomsia and R. O. Ritchie (2009) "Designing highly toughened hybrid composites through nature-inspired hierarchical complexity," Acta Materialia, vol. 57, pp. 2919–2932.

44. R. O. Ritchie, M. J. Buehler and P. Hansma (2009) "Plasticity and toughness in bone," Physics Today, June, pp. 41–47.

45. H. D. Espinosa, J. E. Rim, F. Barthelat and M. J. Buehler (2009) "Merger of structure and material in nacre and bone – Perspectives on de novo biomimetic materials," Progress in Materials Science, vol. 54, pp. 1059–1100.
46. I. Jager and P. Fratzl (2000) "Mineralized collagen fibrils: A mechanical model with a staggered arrangement of mineral particles," Biophysical Journal, vol. 79, pp. 1737–1746.
47. B. Ji and H. Gao (2006) "Elastic properties of nanocomposite structure of bone," Composites Science and Technology, vol. 66, pp. 1212–1218.
48. S. R. Nutt and A. Needleman (1987) "Void nucleation at fiber ends in Al-SiC composites," Scripta Materialia, vol. 21, pp. 705–710.
49. V. Tvergaard (1993) "Model studies of fiber breakage and debonding in a metal reinforced by short fibers," Journal of the Mechanics and Physics of Solids, vol. 41, pp. 1309–1326.
50. M. Finot, Y.-L. Shen, A. Needleman and S. Suresh (1994) "Micromechanical modeling of reinforcement fracture in particle-reinforced metal-matrix composites," Metallurgical and Materials Transactions A, vol. 25A, pp. 2403–2420.
51. S. Ghosh and S. Moorthy (1998) "Particle fracture simulation in non-uniform microstructures of metal-matrix composites," Acta Materialia, vol. 46, pp. 965–982.
52. L. Mishnaevsky, M. Dong, S. Honle and S. Schmauder (1999) "Computational mesomechanics of particle-reinforced composites," Computational Materials Science, vol. 16, pp. 133–143.
53. J. Segurado and J. Llorca (2004) "A new three-dimensional interface finite element to simulate fracture in composites," International Journal of Solids and Structures, vol. 41, pp. 2977–2993.
54. A. Ayyar and N. Chawla (2007) "Microstructure-based modeling of the influence of particle spatial distribution and fracture on crack growth in particle-reinforced composites," Acta Materialia, vol. 55, pp. 6064–6073.
55. E. H. Wong, C. S. Selvanayagam, S. K. W. Seah, W. D. Van Driel, J. F. J. M. Caers, X. J. Zhao, N. Owens, L. C. Tan, D. R. Frear, M. Leoni, Y.-S. Lai and C.-L. Yeh (2008) "Stress–strain characteristics of tin-based solder alloys for drop-impact modeling," Journal of Electronic Materials, vol. 37, pp. 829–836.
56. Abaqus 6.8, User's Manual, Dassault Systèmes Simulia Corp., Providence, RI.
57. Y.-L. Shen and K. Aluru (2010) "Numerical study of ductile failure morphology in solder joints under fast loading conditions," Microelectronics Reliability, doi:10.1016/j.microrel.2010.06.001.
58. A. Hillerborg, M. Modeer and P. E. Petersson (1976) "Analysis of crack formation and crack growth in concrete by means of fracture mechanics and finite elements," Cement and Concrete Research, vol. 6, pp. 773–781.

Index

CPSIA information can be obtained
at www.ICGtesting.com
Printed in the USA
LVOW07*2016080617
537423LV00008B/143/P